VOLUME EIGHT

ADVANCES IN NEUROTOXICOLOGY

Neurotoxicity of Drugs of Abuse

VOLUME EIGHT

ADVANCES IN NEUROTOXICOLOGY

Neurotoxicity of Drugs of Abuse

Series Editors

MICHAEL ASCHNER

LUCIO G. COSTA

Edited by

WILLIAM SLIKKER
Retired, National Center for Toxicological Research/FDA, Jefferson, AR, United States

MICHAEL ASCHNER
Albert Einstein College of Medicine, Bronx, NY, United States

LUCIO G. COSTA
University of Washington, Seattle, WA, United States; University of Parma, Parma PR, Italy

ELSEVIER

ACADEMIC PRESS
An imprint of Elsevier

Academic Press is an imprint of Elsevier
50 Hampshire Street, 5th Floor, Cambridge, MA 02139, United States
525 B Street, Suite 1650, San Diego, CA 92101, United States
The Boulevard, Langford Lane, Kidlington, Oxford OX5 1GB, United Kingdom
125 London Wall, London, EC2Y 5AS, United Kingdom

First edition 2022

ISBN: 978-0-323-91585-4
ISSN: 2468-7480

For information on all Academic Press publications
visit our website at https://www.elsevier.com/books-and-journals

Publisher: Zoe Kruze
Acquisitions Editor: Sam Mahfoudh
Developmental Editor: Jhon Michael Peñano
Production Project Manager: Abdulla Sait
Cover Designer: Miles Hitchen

Typeset by STRAIVE, India

Contents

Contributors

Abdallah Alami
McLaughlin Centre for Population Health Risk Assessment, University of Ottawa; School of Mathematics and Statistics, Carleton University, Ottawa, ON, Canada

G. Caleb Alexander
Center for Drug Safety and Effectiveness, Bloomberg School of Public Health; Division of General Internal Medicine, Johns Hopkins Medicine, Baltimore, MD, United States

G. Jean Harry
Mechanistic Toxicology Branch, Division, National Toxicology Program, National Institute of Environmental Health Sciences, National Institutes of Health, Research Triangle Park, NC, United States

Leah E. Latham
Division of Neurotoxicology, National Center for Toxicological Research/FDA, Jefferson, AR, United States

Edward D. Levin
Professor of Psychiatry and Behavioral Sciences, Duke University Medical Center, Durham, NC, United States

Fang Liu
Division of Neurotoxicology, National Center for Toxicological Research/FDA, Jefferson, AR, United States

Shuliang Liu
Division of Neurotoxicology, National Center for Toxicological Research/FDA, Jefferson, AR, United States

Donald R. Mattison
Department of Epidemiology and Biostatistics, Arnold School of Public Health, University of South Carolina, Columbia, SC, United States; McLaughlin Centre for Population Health Risk Assessment; School of Epidemiology and Public Health, University of Ottawa, Ottawa, ON, Canada

R. Daniel Mellon
Division of Pharmacology/Toxicology for Neuroscience, Office of New Drugs, Center for Drug Evaluation and Research, United States Food and Drug Administration, Silver Spring, MD, United States

Thomas J. Moore
Center for Drug Safety and Effectiveness, Bloomberg School of Public Health, Johns Hopkins University, Baltimore, MD; Department of Epidemiology, Milken Institute School of Public Health, The George Washington University, Washington, DC, United States

Tucker A. Patterson
National Center for Toxicological Research/FDA, Jefferson, AR, United States

Ricardo Marcos Pautassi
Instituto de Investigación Médica M. y M. Ferreyra, INIMEC-CONICET; Facultad de Psicología, Universidad Nacional de Córdoba, Córdoba, Argentina

William Slikker
Retired, National Center for Toxicological Research/FDA, Jefferson, AR, United States

Miriam Beatriz Virgolini
Instituto de Farmacología Experimental de Córdoba-Consejo Nacional de Investigaciones Científicas y Técnicas; Departamento de Farmacología Otto Orsingher, Facultad de Ciencias Químicas, Universidad Nacional de Córdoba, Córdoba, Argentina

Cheng Wang
Division of Neurotoxicology, National Center for Toxicological Research/FDA, Jefferson, AR, United States

Preface

This volume entitled *Neurotoxicity of Drugs of Abuse* provides expert reviews on the disruptive impact of drugs of abuse on the central nervous system. The neurotoxic potential of several agents including marijuana, fentanyl, and ketamine is reviewed in depth, and the consequences of both short-term and chronic use are categorized. Pharmacokinetic profiles as well as mechanisms of action for these and other drugs of abuse such as alcohol and nicotine are also presented. The implications of the abuse of agents such as phencyclidine (PCP) are also discussed.

Important to the study of drugs of abuse is the understanding of the effects of their single and often repeated self-administration, both during adulthood and during development by populations around the world. Abuse liability is seen not only with the agents reviewed in this volume but also with others—including the stimulants—that have been recently reviewed elsewhere (Slikker et al., 2018). One such example is cocaine and the generalization of this drug's propensity for self-administration as it is shared by other stimulant-like drugs including amphetamine and methamphetamine (Slikker et al., 1984). Chronic exposure to cocaine during pregnancy has also been evaluated to determine the impact on maternal and fetal outcomes (Morris et al., 1996).

Among the many questions examined by the authors of this volume is whether the adverse effects of some drugs of abuse on the nervous system are reversible. Sadly, the answer is no in some cases. The authors describe in detail the characteristics of drugs of abuse that produce neurotoxicity in addition to their short- and long-term adverse effects on the nervous system. In addition, the potential mechanisms of action underlying the ability of several centrally acting drugs of abuse to cause neurotoxicity are provided.

Readers including students, basic researchers, clinicians, epidemiologists, regulators, drug developers, and chemists will come away with a better understanding of the disruptive impact of drugs of abuse on the central nervous system. The short- and long-term impacts of several agents including marijuana, fentanyl, ketamine, and PCP are carefully reviewed, along with the most widely abused drugs worldwide: nicotine and alcohol. The implications of short- and long-term abuse of these agents on the central nervous system are fully characterized, providing a state-of-the-art update to the reader.

William Slikker

References

Morris, P., Binienda, Z., Gillam, M.P., Harkey, M.R., Zhou, C., Henderson, G.L., Paule, M.G., 1996. The effect of chronic cocaine exposure during pregnancy on maternal and infant outcomes in the rhesus monkey. Neurotoxicol. Teratol. 18 (2), 147–154.
Slikker Jr., W., Brocco, M.J., Killam Jr., K.F., 1984. Reinstatement of responding maintained by cocaine and thiamylal. J. Pharmacol. Exp. Ther. 228 (1), 43–52.
Slikker Jr., W., Paule, M.G., Wang, C. (Eds.), 2018. Handbook of Developmental Neurotoxicology, second ed. Academic Press.

CHAPTER ONE

Cannabinoids

G. Jean Harry*
Mechanistic Toxicology Branch, Division, National Toxicology Program, National Institute of Environmental Health Sciences, National Institutes of Health, Research Triangle Park, NC, United States
*Corresponding author: e-mail address: harry@niehs.nih.gov

Contents

1. Introduction

Cannabinoids are ligands that are either plant derived, synthetic, or semisynthetic, and have affinity for and activity at cannabinoid receptors. The terms "cannabis" or "marijuana" represent mixtures of dried herbs of the mature female *Cannabis sativa* plant. This plant contains more than 400 different chemical compounds including over 120 different phytocannabinoids (Morales et al., 2017). The most potent of these phytocannabinoids is Δ9-tetrahydrocannabinol (Δ^9-THC). Δ^9-THC functions as a partial agonist for both cannabinoid receptors, CB_1R and CB_2R, and is primarily responsible for the psychoactive and physical effects (Howlett et al., 2002). Besides Δ^9-THC, other cannabinoids are present at lower levels including Δ8-tetrahydrocannabinol (Δ^8-THC), cannabinol (CBN), cannabidiol (CBD), cannabigerol (CBG), cannabichromene

Advances in Neurotoxicology, Volume 8
ISSN 2468-7480
https://doi.org/10.1016/bs.ant.2022.06.001

(CBC), Δ^9-tetrahydrocannabivarin (THCV), cannabivarin (CBV), and cannabidivarin (CBDV) (Pertwee, 2005). Δ^8-THC is slightly less potent than Δ^9-THC and is present in very small quantities in the plant. CBD lacks psychoactive effects, but has been shown to act as an anti-inflammatory, antioxidant, anti-epileptic, antirheumatic, anxiolytic and analgesic drug (Ashton, 2001; Britch et al., 2020; Dos Santos et al., 2021; Tham et al., 2019). Many of the phytocannabinoids have partial agonist activity at cannabinoid receptors and potentially modulate Δ^9-THC effects through additive, synergistic, or antagonist activity (Russo and Guy, 2006; Solymosi and Köfalvi, 2017; Zagzoog et al., 2020).

Synthetic cannabinoids (SCs) are a collection of diverse compounds that exhibit high affinity binding to the CB_1R and CB_2R and display a similar pharmacological profile to that of THC (Ford et al., 2017; Potts et al., 2020; Worob and Wenthur, 2019). Synthetic cannabinoids have been classified as classical, non-classical, and aminoalkylindole based on their chemical structure. They have different levels of intrinsic activity and selectivity for cannabinoid receptors (Pertwee et al., 2010). The classical group consists of dibenzopyran derivatives of THC including HU-210, HU-243, and nabilone. The non-classical group consists of bi-cyclic and tri-cyclic analogs of THC that lack the pyran ring. The aminoalkylindole group is of a completely different structure as compared to THC (Pertwee et al., 2010). Most synthetic cannabinoids are highly lipophilic allowing for easy access across the blood–brain barrier. The higher efficacy and potency at cannabinoid receptors plus non-cannabinoid receptor sites result in a potentially greater health threat from illicit synthetic cannabinoids than Cannabis/THC (Gerostamoulos et al., 2015; Grim et al., 2016). Synthetic cannabinoids that are indole-derivatives or members of the aminoalkylindole family such as JWH-073 (1-butyl-1H-indol-3-yl)-1-naphthalenyl-methanone) have fourfold higher binding affinity toward central CB_1Rs than Δ^9-THC (Aung et al., 2000; Brents et al., 2012; Wiley et al., 1998). JWH-073 is biotransformed in vivo into mono-hydroxylated metabolites retaining significant affinity and activity at CB_1Rs (Brents et al., 2012). JWH-210 has $20\times$ high binding affinity toward central CB_1Rs as compared JWH-073 and $100\times$ higher than Δ^9-THC. In rodents, each of these synthetic cannabinoids show activity at CB_1Rs producing "tetrad" effects of Δ^9-THC such as hypothermia, antinociception, suppressed locomotor activity, akinesia (Brents et al., 2012; Marshell et al., 2014; Wiley et al., 1998).

The biological effects of phytocannabinoids or synthetic cannabinoids, like endogenous cannabinoids, occur through interactions with the endocannabinoid system (Araque et al., 2017; Gamage et al., 2018; Howlett and Abood, 2017). This system is essential for a diverse range of brain functions including mood balance, energy balance, learning and memory and has been implicated in multiple psychiatric, neurological, neurodegenerative, and neuroinflammatory related diseases. Positive effects on mood are associated with the use of cannabis, but there is also the potential slowed reaction time, motor incoordination, short-term memory loss, and attention disruptions (Bourque and Potvin, 2021; Brucki et al., 2021; Pratt et al., 2019). Cannabinoids may also affect the cardiovascular and respiratory systems and have immunosuppressive and endocrine properties (Solymosi and Köfalvi, 2017). In humans, nausea, seizures, sensorimotor responses, hallucinations, altered cognitive abilities, and cardiovascular impairment have been reported (Doğan et al., 2016; Hermanns-Clausen et al., 2016; Ossato et al., 2016; Tait et al., 2016; Zawilska and Wojcieszak, 2014). The effects of cannabis are influenced by the relative content in different phytocannabinoids, route of administration, length of use, polyconsumption, drug history, and age of subject (Bara et al., 2021; Higuera-Matas et al., 2015; Zamberletti and Rubino, 2022). There is a paucity of detailed examinations of CBD formulations, however, clinically they have been reported to have diverse medicinal properties that can be accompanied by side effects such as fatigue, malaise, sleep disorders, and infections (Pauli et al., 2020).

2. The endocannabinoid system (ECS)

The endogenous cannabinoid system (ECS) is a lipid-based signaling system comprised of ligands that bind to cannabinoid receptors (Lu and Mackie, 2016; Rodríguez de Fonseca et al., 2005; Schurman et al., 2020). Components of the ECS include N-arachidonoylethanolamine (anandamide, AEA), 2-arachidonoglycerol (2-AG), synthetizing and degrading enzymes, receptors (e.g., CB_1Rs and CB_2Rs), endocannabinoid membrane transporter (EMT), and CB_1R regulator interacting protein 1a (CRIP1a). Endocannabinoid biosynthesis occurs "on demand" from phospholipids at the plasma membrane in response to increased intracellular Ca^{2+} or activation of the phospholipase C (PLC) pathway. Ligands are released from the postsynaptic neurons and act within a 20 μm area site (Wilson and Nicoll, 2002) with a <5 min half-life (Willoughby et al., 1997). These ligands act at the presynaptic receptor via retrograde transmission to dampen presynaptic neurotransmitter release serving as a sensor-feedback mechanism to modulate synaptic activity (Freund et al., 2003; Hillard, 2015; Yoshida et al., 2006).

2.1 Endogenous cannabinoid receptor ligands

Endocannabinoids are members of a family of bioactive lipids that are derivatives of arachidonic acid conjugated with ethanolamine or glycerol (Di Marzo, 2018). Precursors of endocannabinoids are present in lipid membranes and are liberated following activation of certain G protein-coupled receptors or by depolarization. This occurs by rapid enzymatic steps for release of ligands into the extracellular space. Four primary enzymes have been characterized: (1) AEA; (2) 2-AG, the ester of arachidonic acid and ethanolamine; (3) virodhamine, which resembles AEA; and (4) noladim, an analogue of 2-AG, 2-arachidonyl glyceryl ether. AEA and 2-AG are primary enzymes and with well characterized synthesis through enzymatic conversion of membrane lipids (Hillard, 2015; Maccarrone, 2009; Sugiura et al., 2006). 2-AG is a metabolic intermediate in lipid metabolism highly concentrated in the central nervous system (CNS). AEA is the product of the cleavage of a membrane phospholipid and has a CNS concentration 200-fold lower than 2-AG, however, this can transiently increase by 5- to 12-fold after depolarization or receptor stimulation (Stella and Piomelli, 2001; Kim et al., 2002).

N-arachidonoyl-phosphatidylethanolamine (NAPE), the principal phospholipid precursor for AEA synthesis, is synthesized by the calcium-dependent enzyme, N-acyltransferase (NAT), and catalyzes the transfer of arachidonic acid from phosphatidylcholine to the head group of phosphatidylethanolamine. NAT requires the presence of Ca^{2+} and is regulated by cAMP. Phosphorylation of NAT is mediated through cAMP-dependent activity of protein kinase A (Cadas et al., 1996; Piomelli, 2003). The resulting N-arachidonoyl-phosphatidyl ethanolamine (NArPE) is cleaved by a calcium-dependent NAPE (N-acyl-phosphatidylethanolamine)-specific phospholipase D (NAPE-PLD). This leads to the release of AEA from membrane lipids. While this is the primary route for AEA synthesis, other routes have been proposed including: (1) dual hydrolysis of acyl groups by the phospholipase B, α,β hydrolase domain containing 4, N-acyl phospholipase B (ABHD4), followed by hydrolysis by the glycero-phospho B-acyl ethanolamine phosphodiesterase, GDE1 (Simon and Cravatt, 2010), and hydrolysis of one acyl group; (2) AEA liberation by a lyso-NAPE-PLD (Tsuboi et al., 2013); and (3) NAPE-phospholipase C (PLC) followed by phosphatase. The NAPE-PLC is a route found primarily in immune cells (Liu et al., 2006).

2-AG is a monoglyceride that can be synthesized by two different metabolic pathways. It is produced from sequential hydrolysis of 2-arachidonoyl-containing phospholipids, primarily arachidonoyl-containing phosphatidyl inositol bis-phosphate (PIP2), by PLCβ. This is followed by hydrolysis of the resulting diacylglycerol (DAG) by diacylglycerol lipase (DAGL) and the release of DAG from membrane phospholipids by PLC (Bisogno et al., 2003; Murataeva et al., 2014; Tsuboi et al., 2018). There are two isoforms of DAGL (Bisogno et al., 2003). In the adult CNS, DAGLα is dominant and responsible for most 2-AG production contributing to synaptic plasticity (Gao et al., 2010; Tanimura et al., 2010). Anatomical studies have identified a proximity of DAGLα receptors and metabotropic glutamatergic receptor subtype 5 (mGluR5) in dendritic spines and apposed to presynaptic CB1 receptors (Katona et al., 2006; Yoshida et al., 2006). In contrast, DAGLβ plays an important role in the generation of 2-AG during immune responses (Hsu et al., 2012). However, when cooperation with DAGLα is required, DAGLβ contributes to synaptic 2-AG (Jain et al., 2013). A secondary pathway for 2-AG synthesis is the cleavage of the phosphatidyl inositol precursor by phospholipase A followed by hydrolysis of the phosphate ester bond by lyso-phospholipase C. Hydrolyzation by monoacylglycerol lipase (MAGL) produces arachidonic acid and glycerol (Maccarrone, 2009). Additional 2-AG biosynthetic pathways include PLA1 activation of lysophospholipase C (lyso-PLC) and dephosphorylation of arachidonoyl-lysophosphatidic acid (Nakane et al., 2002).

AEA and 2-AG are not stored in vesicles but are synthetized and released "on demand." Different enzymatic steps can lead to "on demand" AEA cleavage from NAPE. These steps can involve the Ca^{2+}-dependent activation of specific phospholipases A2, C, D; and α, β domain serine hydrolase 4 (ABDH4). Regulation of phospholipase D (PLD) activity occurs by depolarization or by activation of the ionotropic glutamate *N*-methyl-D-Aspartate (NMDA) receptors; nicotinic α7 neuronal receptors (Piomelli, 2003; Stella and Piomelli, 2001); or metabotropic receptors of acetylcholine, dopamine, and glutamate (Giuffrida et al., 1999; Kim et al., 2002; Varma et al., 2001). Production of 2-AG can be induced by the activation of PLC upon stimulation of group I metabotropic glutamate,

M1 or M3 muscarinic, and orexin-A receptors (Piomelli, 2003; Stella et al., 1997). Activation of ionotropic purinergic receptors can boost 2-AG formation (Witting et al., 2004). Once AEA and 2-AG are formed, they target CB_1Rs in the same originating cell via diffusion within the plasmalemma or they can be released to the extracellular fluid where they reach presynaptic terminals. This occurs with help of protein carriers such as lipocalins or albumin (Piomelli, 2003). NAPE-phospholipase D is localized predominantly in presynaptic sites of hippocampal excitatory afferents. The highest density of NAPE-PLD is found in CB_1R-negative mossy fiber terminals (Nyilas et al., 2008). In the ventral pallidum, NAPE-PLD is presynaptic in both inhibitory and excitatory terminals that are predominantly CB_1R-negative and appose CB1R-positive nerve terminals (Pickel et al., 2012). In dendrites, DAGLα is induced to produce 2-AG upon depolarization-induced Ca^{2+} entry and/or stimulation of Gq/11 protein-coupled metabotropic receptors (Castillo et al., 2012). An increase in 2-AG signaling through CB_1Rs is often accompanied by COX-1 inhibition (Kim and Alger, 2004; Straiker et al., 2011). Generally, 2-AG, acting via CB_1Rs, suppresses synaptic transmission and neuronal excitability. However, a major oxidative metabolite of 2-AG, prostaglandin E2-glycerol ester (PGE2-GE), potentiates synaptic transmission, enhances synaptic plasticity, and produces hyperalgesia (Hu et al., 2008; Sang et al., 2006; Yang et al., 2008). Microglial cells and astrocytes also express the AEA-synthesizing enzyme NAPE-PLD (Kallendrusch et al., 2012) and AEA (Gabrielli et al., 2015; Walter et al., 2002). The production of AEA by microglia and release of the endocannabinoid to the perivascular space occurs in association with macrovesicles or exosomes serving to modulate the activity of inhibitory neurons (Gabrielli et al., 2015). FAAH and fatty-acid binding proteins (FABPs), which participate in the transport and hydrolysis of AEA, respectively, are expressed in microglia, astrocytes, oligodendrocytes, and NG2-positive cells (Egertová et al., 2003; Young et al., 2013; Duffy et al., 2017; Gerstner et al., 2017; Foerster et al., 2020).

Endocannabinoids can also serve as retrograde messengers for neuronal depolarization (Ohno-Shosaku et al., 2001; Wilson and Nicoll, 2002) and contribute to short-term synaptic plasticity (Alger, 2012; Castillo et al., 2012; Kano et al., 2009; Neu et al., 2007; Safo et al., 2006). Endocannabinoids can affect synaptic plasticity by (1) depolarization-induced suppression of inhibition (DSI)/depolarization-induced suppression of excitation (DSE); (2) metabotropic-induced suppression of inhibition (MSI)/metabotropic-induced suppression of excitation or endocannabinoid-mediated short-term depression; or (3) endocannabinoid-mediated long-term depression (eCB-LTD). LTD is a process representing a long-lasting inhibition of synaptic strength. Endocannabinoids can induce both homosynaptic LTD at the stimulated synapse and heterosynaptic LTD at synapses adjacent to the stimulated synapses. The loss of inhibition increases dendritic excitability (Chevaleyre and Castillo, 2004) and modifies function and maturation of neural circuits (Jiang et al., 2010). Endocannabinoids can directly inhibit neuronal excitability through CB_1Rs (Chávez et al., 2010; Kellogg et al., 2009; Marinelli et al., 2008) or slow-self inhibition (Bacci et al., 2004). This is observed in low threshold-spiking cortical interneurons and cerebellar basket cells (Bacci et al., 2004; Kreitzer et al., 2002) and, possibly in cortical neurons (Marinelli et al., 2009).

2.2 Endocannabinoid signaling termination and degradation

Endocannabinoid signaling is terminated by a two-step process that includes transport into cells and hydrolysis by two specific enzymatic systems (Fegley et al., 2004). Endocannabinoid uptake is mediated by a transporter widely distributed throughout the brain and facilitates the uptake of both AEA and 2-AG in an energy-independent fashion (Beltramo et al., 1997; Giuffrida et al., 2001). The AEA transporter is saturable, displays substrate specificity, and can be blocked by specific drugs. The degradation of endocannabinoids is accomplished by the fatty acid amide hydrolase (FAAH) (Cravatt et al., 1996) and the monoacylglyceride lipase (MAGL) (Dinh et al., 2002). The main enzyme responsible for 2-AG inactivation is MAGL (Blankman et al., 2007; Dinh et al., 2002), which is a serine hydroxylase highly expressed in presynaptic terminals in the central nervous system (Gulyas et al., 2004; Ludanyi et al., 2011; Matyas et al., 2006; Uchigashima et al., 2007). Degradation of AEA occurs through (1) FAAH (Cravatt et al., 1996); (2) oxidation by cyclooxygenase-2 (COX-2) to create prostamides (prostagladin-ethanolamides) (Di Marzo, 2008; Woodward et al., 2008); and (3) N-acylethanolamine-hydrolyzing acid amidase (NAAA) (Tsuboi et al., 2005). For 2-AG, the major degradative pathway is via hydrolysis to arachidonic acid and glycerol (Di Marzo, 2008). Degradation is primarily due to the hydrolytic enzyme, MAGL and, to a lesser extent ($<10\%$), by alpha/beta domain hydrolases (ABHD) 6 and 12 (ABHD6 and 12) (Blankman et al., 2007). Additionally, 2-AG can be oxidized by COX-2 (Woodward et al., 2008) and hydrolyzed by FAAH transformation (Hermanson et al., 2014; Kozak et al., 2000), cytochrome P450 (Chen et al., 2008), and lipoxygenases (Maccarrone et al., 2000). A central role of MAGL in 2-AG degradation in mice is evident with pharmacological inhibition of MAGL (Long et al., 2009). MAGL inhibition increases 2-AG signaling at CB_1Rs. It reduces available levels of arachidonic acid necessary for prostaglandin synthesis and thus, dampens processes related to inflammatory responses (Schlosburg et al., 2010; Ignatowska-Jankowska et al., 2014; Kinsey et al., 2011, 2013; Nomura et al., 2011).

ABHD6 is a postsynaptic integral membrane protein with an intracellular-facing active side. It is primarily localized on the dendrites and dendritic spines of excitatory neurons in the cortex (Marrs et al., 2010). Inhibition of ABHD6 increases 2-AG signaling through CB_1Rs (Marrs et al., 2010; Naydenov et al., 2014). This leads to a suppression of synaptic transmission and neuronal excitability contributing to CB_1R-dependent long-term depression (LTD). In mouse microglia, ABHD6 controls the accumulation of 2-AG and the knockdown of ABHD6 increases the efficacy with which 2-AG can stimulate CB_2R-mediated cell migration (Marrs et al., 2010). ABHD12 is a major brain lysophosphatidylserine (lyso-PS) suggested to be involved in the hydrolysis of 2-AG. It plays a significant role in the degradation of long chain lyso-phosphatidylserines (lyso-PSs) (Blankman et al., 2013). Loss-of-function mutations of ABHD12 in humans is associated with increased levels of lyso-PSs (Kamat et al., 2015) and the neurodegenerative polyneuropathy, hearing loss, ataxia, retinitis pigmentosa, and cataract (PHARC) syndrome (Fiskerstrand et al., 2010; Nguyen et al., 2021).

2.3 Cannabinoid receptors

Cannabinoids function primarily through activation at CB_1Rs and CB_2Rs. The CB_1R has seven transmembrane domains and two terminal tails, one N-terminal in the extracellular part and one intracellular C-terminal part, plus three extracellular and three intracellular loops. CB_2R is also a member of the G protein-coupled family with seven transmembrane domains, along with an extracellular N-terminus, an intracellular C-terminus, three extracellular and three intracellular loops (Jordan and Xi, 2019; Li et al., 2019). Callén et al. (2012) demonstrated molecular and functional heteromerization of CB_1Rs and CB_2Rs in a variety of brain regions suggesting receptor interactions. Individual endocannabinoids exhibit different binding properties and intrinsic activity at CB_1Rs and CB_2Rs. The endogenous cannabinoids, AEA and 2-AG, bind to the CB_1R orthosteric site as the main active site (Kendall and Yudowski, 2016). AEA behaves as a partial agonist at both CB_1Rs and CB_2Rs showing a higher affinity for the CB_1R (Hillard et al., 1999; Howlett et al., 2002). The intrinsic activity of AES at CB1Rs is 4- to 30-fold higher than activity at CB_2Rs. 2-AG is a complete agonist at both receptors and exhibits less receptor affinity than AEA (Howlett et al., 2002; Stella et al., 1997). Allosteric modulators that bind to CB_1R sites different from the orthosteric sites have been reported [e.g., Org27569 (Price et al., 2005), PSNCBAM-1 (Horswill et al., 2007), RTI-371 (Navarro et al., 2009), and ZCZ011 (Ignatowska-Jankowska et al., 2015)]. These allosteric modulators may reflect novel mechanisms of action from ligand binding to cytosolic signal transduction. Both orthosteric (Lauckner et al., 2005; Mukhopadhyay and Howlett, 2005) and allosteric (Khurana et al., 2017) ligands can induce a preference for G-protein coupling and downstream signaling.

The psychoactive and pharmacological properties of THC have been associated with agonist activities at both receptors (Cox et al., 2007; Di Marzo et al., 2000; Laaris et al., 2010; Ledent et al., 1999), while effects on the immune system appear to be related to activity on CB_2Rs or their ability to activate adenosine type 1 receptors (Gonca and Darıcı, 2015). The cannabinoid ligands have their biological function through G-protein-coupled membrane receptors, which primarily couple to G proteins of the Gi and Go classes (Matsuda et al., 1990; Munro et al., 1993). There is some evidence that CB_1R but not CB_2R can associate with Gs (Calandra et al., 1999). Stimulation of either receptor has been shown to activate ERK1/2, phosphoinositide 3-kinase and protein kinase C (PKC) by activating Gi/o proteins, inhibiting adenylyl cyclase, and altering protein kinase activity (Davis et al., 2003).

2.3.1 Cannabinoid 1 receptors (CB_1Rs)

CB_1R expression is observed in the testis, some glandular systems, the microcirculation, and the nervous system (Batkai et al., 2001; Herkenham et al., 1991; Wagner et al., 1997). Human, rat, and mouse CB_1Rs show 97–99% amino acid sequence identity and similar pharmacological properties but some variation in ligand binding (McPartland et al., 2007). In the human brain, CB_1R densities are greater in regions associated with cognitive function (frontal cortex) and memory (hippocampus), while in rodents they are greater in movement-associated areas (cerebellum) and the caudate-putamen (Herkenham et al., 1990). CB_1Rs show a greater distribution in the human dorsomedial nucleus, an area that mediates emotions from the amygdaloid complex and contributes to personality formation. In addition to distribution levels, a more complex CB_1R

architecture has been observed in primates as compared to rodents in the cingulate and frontal cortex, hippocampus, amygdala (emotional behaviors), and parietotemporal areas (visual and auditory information processing) (Harkany et al., 2005). Roles have been identified for neuronal CB_1Rs signal transduction in the regulation of neurotransmission (Kano, 2014; Lu and Mackie, 2016), neurodevelopment (Díaz-Alonso et al., 2012a,b; Gaffuri et al., 2012;Maccarrone et al., 2014), and synaptic plasticity (Araque et al., 2017; Garcia et al., 2016).

Relative to classical neurotransmitters such as dopamine or opioid receptors, the CB_1R-G-protein-coupled receptor (GPCR) shows 10- to 50-fold higher densities (Herkenham et al., 1991). Another important characteristic of CB_1R is the low efficiency coupling to its transduction system (Breivogel et al., 1998; Felder and Glass, 1998). CB_1R agonists promote dissociation of the G protein from the CB_1R, whereas antagonist and inverse agonists maintain the CB_1R-Gi protein interaction in a more stable form (Mukhopadhyay and Howlett, 2005). Following Gi/o activation and dissociation, CB_1Rs are phosphorylated by G-protein receptor kinases (GRKs) for interactions with the scaffolding proteins β-arrestin 1 or β-arrestin 2 (Breivogel et al., 2013; Chen et al., 2014). These then internalize CB_1Rs from the plasma membrane to regulate non-G protein CB_1R-mediated signal transduction (Nogueras-Ortiz and Yudowski, 2016), such as extracellular signal-regulated kinase (ERK)1 and 2 phosphorylation and activation (Daigle et al., 2008; Dalton and Howlett, 2012; Rubino et al., 2006).

CB_1Rs are present in all plasma membrane components including lipid rafts (Asimaki and Mangoura, 2011; Bari et al., 2005; Barnett-Norris et al., 2005; Sarnataro et al., 2005). The expression at the plasma membrane comprises only $\sim$20% of total CB_1Rs (Leterrier et al., 2006; McIntosh et al., 1998; Rozenfeld and Devi, 2008) with the majority observed intracellular. In neurons, the plasma membrane surface expression is highly polarized toward the axonal compartment following a disto-proximal gradient (Fletcher-Jones et al., 2020; Simon et al., 2013; Wickert et al., 2018). Plasma membrane CB_1Rs are primarily located on presynaptic terminals (Mackie, 2005; Marsicano and Lutz, 1999; Matyas et al., 2006; Nyiri et al., 2005; Tsou et al., 1998) of both GABAergic (γ-aminobutyric acid) and glutamatergic neurons where they modulate neurotransmitter release (Katona et al., 1999, 2006; Kawamura et al., 2006). This occurs primarily from GABAergic neurons and to a lesser degree from glutaminergic neurons (Katona et al., 1999; Puighermanal et al., 2012; Walsh and Andersen, 2020). Functional subpopulations of CB_1Rs are also observed at the surface of soma and dendrites (Bacci et al., 2004; Marinelli et al., 2009). CB_1Rs in the cortex and hippocampus are enriched on cholecystokinin (CCK) positive interneurons (Bodor et al., 2005; Katona et al., 1999; Tsou et al., 1999) and on medium spiny striatal neurons in the dorsal and ventral striatum (Matyas et al., 2006, 2007;Uchigashima et al., 2007). High receptor expression is seen on the direct pathway axons entering the globus pallidus (Tsou et al., 1998). In the cerebellum, CB_1Rs are found in the parallel and climbing fibers and in basket cells (Safo et al., 2006; Suarez et al., 2008).

Glial elements (astrocytes, microglia, oligodendrocytes) show functionally relevant low-level expression of CB_1Rs (Han et al., 2012; Ilyasov et al., 2018; Mecha et al., 2015; Molina-Holgado et al., 2002; Oliveira da Cruz et al., 2016; Stella, 2010). In astrocytes, activation of CB_1Rs can modulate cellular metabolism (Jimenez-Blasco et al., 2020), mediate neuron-astrocyte communication (Navarrete and Araque, 2008), and contribute to synaptic and memory impairments (Chen et al., 2013) and neurological disorders (Scarante et al., 2021). In addition, CB_1R activation on astrocytes contributes to the modulation of neuroinflammation through interleukin-1 (IL-1) (Fields et al., 2022) or cyclooxygenase-2 (COX-2) (Cong et al., 2021; Zhang and Chen, 2008).

Intracellular CB_1Rs are associated with endosomes and mitochondria (Benard et al., 2012; Djeungoue-Petga and Hebert-Chatelain, 2017; Rozenfeld and Devi, 2008; Xu et al., 2016). Activation of CB_1Rs associated with mitochondrial outer membranes hinders glucose metabolism and lactate production (i.e., mitochondrial respiration) (Benard et al., 2012; Jimenez-Blasco et al., 2020; Koch et al., 2015) and is implicated in certain types of memory (Hebert-Chatelain et al., 2016). Internalization of CB_1Rs is dynamically regulated by receptor stimulation primarily due to receptor phosphorylation and β-arrestin binding followed by clathrin-mediated endocytosis (CME) (Ahn et al., 2013; Blume et al., 2016) or alternatively by caveolin-mediated endocytosis of proteins localized to lipid rafts (Razani et al., 2002). At these locations, the receptors mediate slow self-inhibition. It is considered that the differential rates of endocytosis between axons and dendrites contributes to maintaining CB_1R polarity.

Most GPCRs are endocytosed from the cell surface following activation. The receptor is then either recycled back to the surface or trafficked for degradation (Ferguson, 2001). The available data suggest that recycling of the receptor is preferred following a relatively short time-period of agonist exposure (<20 min) or at low agonist concentrations (Blume et al., 2016; Grimsey et al., 2010; Tappe-Theodor et al., 2007). With prolonged exposure or at higher concentrations of agonist, there is a reduction of surface levels and an accumulation in lysosomal compartments with recovery requiring synthesis of new receptors. The intracellular sorting of CB_1R to lysosomes following chronic agonist exposure has been linked to the G protein-coupled receptor associated protein 1 (Simonin et al., 2004). Prolonged agonist exposure and subsequent downregulation and desensitization of CB_1R along with changes in receptor G protein coupling and endocannabinoid content, have been implicated in the development of cannabis tolerance and dependence (Henderson-Redmond et al., 2020; Hsieh et al., 1999; Martini et al., 2010; Mato et al., 2005; Morgan et al., 2014; Nealon et al., 2019; Wu et al., 2008).

2.3.2 Cannabinoid 2 receptor (CB_2R)

CB_2R shares ~45% amino acid homology with the CB_1R. CB_2R is coupled predominantly to Gi/o proteins with linkage to signaling cascades involving adenylyl cyclase and cAMP, mitogen-activated protein kinase (MAPK), and intracellular calcium regulation (Chen et al., 2017; Onaivi et al., 2006). An extensive characterization of a panel of ligands binding to CB_2R s showed evidence of agonism with respect to GTPγS, cAMP, β-arrestin, pMAPKs, and G-protein-gated inwardly rectifying potassium channel (Soethoudt et al., 2017). Activation of CB_2R s has been shown to reduce translocation of NF-κB p65 to the nucleus, perhaps as a downstream consequence of MAPK signaling (Correa et al., 2010; Javed et al., 2016). Cannabinoid receptors have been found to form oligomeric receptor complexes, known as heteromers, with several other class A GPCRs. In the nervous system, CB2-A2A, CB2-5HT1A, and CB2-CB1 heteromers have been observed within microglia and in different quantities under conditions of neuroinflammation and in response to different cannabinoid treatments (Bagher et al., 2020; Navarro et al., 2018). Within CB_2R heteromers, allosteric control can occur via cross-antagonism to diminish signaling from the CB_2Rs upon antagonism of the partner receptor. This is often a bidirectional phenomenon where a CB_2R antagonist would also serve as an antagonist for the partner receptor. Thus, upon negative allosteric control a reduced signal occurs from the CB_2Rs relative to the non-heteromeric complex receptor (Callén et al., 2012; Franco et al., 2019). In addition to CB_1Rs and CB_2Rs, there is evidence of non-CB_1Rs and non-CB_2Rs mediating cannabinoid effects (Pertwee et al., 2010). Candidate receptors include the transient receptor potential cation channel subfamily V member 1 (capsaicin (TRPV1) or vanilloid receptor VR1), peroxisome proliferator activated receptors, and G-protein-coupled receptors GPR55 and GPR119.6 (Brown, 2007; Irving et al., 2017; O'Sullivan, 2007).

The CB_2R is present in multiple lymphoid organs with high expression in B lymphocytes, moderate expression in monocytes and polymorphonuclear neutrophils, and low expression in T lymphocytes and elements of the vascular system, where they may modulate immune-related responses (Carlisle et al., 2002; Galiègue et al., 1995; Lu and Mackie, 2016; Munro et al., 1993; Piomelli, 2003; Tsuboi et al., 2018). As compared to CB_1Rs, CB_2Rs are expressed at much lower levels in the CNS with a primary localization to microglia (Atwood and Mackie, 2010). A low-level receptor expression is observed in the normal brain tissue including in the cortex, hippocampus, cerebellum, ventral tegmental area, and brain stem (Gong et al., 2006; Li and Kim, 2015). In the human brain, CB_2Rs have been detected on perivascular microglia (Nunez et al., 2004). Examination of microglia in culture indicated a localization of CB_2Rs to the leading edge of microglial lamellipodia suggesting a role for the receptor in cell migration to a site of injury or stimulus (Walter et al., 2003). However, in human monocytes the activation of CB_2Rs by JWH-015 showed reduced migration (Montecucco et al., 2008). In the murine microglia BV-2 cell line, ABHD6 controls the accumulation of 2-AG and, in the absence of ABHD6, there is an increased efficacy with which 2-AG can stimulate CB_2R-mediated cell migration (Marrs et al., 2010). There has been little evidence that neurons express CB_2Rs under normal conditions with a demonstrated low-level expression in midbrain dopaminergic neurons in the ventral tegmental area, the prefrontal cortex, cerebellum, dorsal striatum, and nucleus accumbens (Zhang et al., 2014, 2017); however, upon injury, specific neurons may display functional receptors (Van Sickle et al., 2005; Viscomi et al., 2009).

The presence on immune cells allows CB_2Rs to modulate responses to inflammatory stimuli (Buckley et al., 2000; Maresz et al., 2005; Ramirez et al., 2012; Walter et al., 2003). CB_2Rs can be induced in the CNS in response to injury or pathophysiological states (Guindon and Hohmann, 2009; Jordan and Xi, 2019; Roche and Finn, 2010; Woodhams et al., 2017). Enhanced expression on microglia or infiltrating myeloid cells has been observed in various models of neurological disorders that are normally associated with an inflammatory environment such as Alzheimer's disease (Benito et al., 2003; Horti et al., 2010; Savonenko et al., 2015; Solas et al., 2013).

In models of pain, an association has been demonstrated with CB_2R expression on neurons, lymphocytes, and/or microglia/macrophages (Cabañero et al., 2020). In comparison, in Huntington's disease, immunoreactivity is localized to blood vessel endothelium and vascular smooth muscle rather than in microglia in the human (Dowie et al., 2014) but can be observed in microglia in an animal model (Sagredo et al., 2009).

The localization on CB_2Rs on immune cells has suggested that they may serve as potential targets for developing therapeutical approaches to address inflammatory-related responses associated with disease in the absence of cannabimimetic effects (Aly and Masocha, 2021; Donvito et al., 2018; Ferrisi et al., 2021; Han et al., 2014; Hashiesh et al., 2021; Leleu-Chavain et al., 2013; Ligresti et al., 2016). While there is a current interest in the targeting of CB_2Rs for treating pain, the recent literature review of clinical trials evaluating CB_2R agonist analgesic effects in chronic pain showed a track record of failure (Cabanero et al., 2021). While pharmacological activation of CB_2Rs appears to not directly induce a pro-inflammatory or anti-inflammatory response of microglia, an attenuation of a stimulus-induced pro-inflammation response has been observed (Correa et al., 2010; Ehrhart et al., 2005; Ma et al., 2015; Ramirez et al., 2005). Additionally, activation of CB_2Rs can contribute to the regulation of microglia phagocytosis (Ehrhart et al., 2005). The directionality of the response appears to be related to the nature of the stimulus. For example, CB_2Rs are upregulated in cultured microglia following stimulation with interferon gamma (IFNγ) plus the monocyte growth factor, granulocyte-macrophage colony-stimulating factor (Maresz et al., 2005). When challenged with lipopolysaccharide (LPS) or LPS/IFNγ, cultured microglia decrease their expression of CB_2Rs (Maresz et al., 2005; Schmöle et al., 2015a,b). Data from CB_2R deficient mouse strains show less than uniform responses with diminished or absent responses to known pro-inflammatory and anti-inflammatory stimuli (Mecha et al., 2015; Schmöle et al., 2015a,b) suggesting a more complex biological outcome than simply anti-inflammation. This complexity is reflective in the work of Liu et al. (2020) showing an involvement of CBRs in microglia and dopaminergic neurons in the "tetrad" effects of the mixed agonist WIN 55212-2 and the selective agonists for the CB_1Rs and CB_2Rs, ACEA and JWH133, respectively.

Selective activation of CB_2Rs on cultured microglia by the synthetic cannabinoid, JWH-015, suppresses CD40 expression induced by IFNγ (Ehrhart et al., 2005). In a transgenic mouse model of Alzehimer's disease, oral administration of JWH-133 to stimulate CB_2Rs resulted in lower levels of Aβ in the cortex accompanied by lower levels of COX-2 and tumor necrosis factor α (TNFα) mRNA expression (Martin-Moreno et al., 2011). This would be consistent with the observation that activation of CB_2Rs enhanced phagocytic engulfment of amyloid β (Aso et al., 2016; Ehrhart et al., 2005; Tolón et al., 2009). A similar conclusion of an anti-inflammatory effect of CBD and WIN 55,212-2 was reported by Martin-Moreno et al. (2011) with a decrease in Aβ-induced IL-6. In comparison, administration of the selective CB_2R agonist, JWH-133, resulted in a reduction in TNFα, IL-1β, IL-6 in amyloid precursor protein/presenilin 1 (APP/PS1) mice (Aso et al., 2013). Additional experiments showed a higher percentage of microglia expressing IL-6 and IL-10 in APP/PS1 mice which was interpreted as representative of immunoregulatory activity (Aso et al., 2013). Within the framework of inflammaging, the elevation of a pro-inflammatory state with aging, WIN 55,212-2 delivered to aged animals resulted in a decrease in TNFα and IL-1β (Marchalant et al., 2009). Similar beneficial effects of CB_2R activation have been observed in animal models of dopaminergic neuronal damage (Javed et al., 2016; Ojha et al., 2016; Price et al., 2009) and multiple sclerosis. In the experimental autoimmune encephalomyelitis (EAE) mouse model (Constantinescu et al., 2011), CBD inhibited the infiltration of inflammatory immune cells into the CNS (Giacoppo et al., 2017). This was accompanied by phosphorylation of PI3K, Akt, and mTOR and the inhibition of MAPK signaling to induce an anti-inflammatory response (Giacoppo et al., 2017). In human and murine NSCs, the activation of GPR55 was neuroprotective against the effects of IL-1β or LPS (Hill et al., 2019).

3. Endocannabinoid system influence on the generation of neural cells

3.1 Developmental

Signaling via the endocannabinoid system is an essential component in the generation of neural cells during organogenesis and in the neurogenic regions of the mature brain. Endocannabinoid system signaling is active in ventricular/subventricular zones from murine embryonic day (ED) 12 onwards where it serves to regulate proliferation of neuronal progenitors (Aguado et al., 2005; Mulder et al., 2008). In the embryo, CB_1R mRNA is expressed in the neural tube and in several distinct structures within the central nervous system by ED15 to ED21. High levels of CB_1R mRNA are also present in the sympathetic and parasympathetic ganglia, retina and, enteric ganglia of the gastrointestinal tract (Buckley et al., 1998). A similar early protein expression is seen in the brain for endocannabinoids and cannabinoid receptors (Berrendero et al., 1999; Fernández-Ruiz et al., 2000). Signaling within the endocannabinoid system serves to regulate neural development showing effects on neural progenitor proliferation, lineage commitment, radial and tangential migration of pyramidal cortical neuron and interneurons, axonal guidance, neuronal maturation, and synaptogenesis (Galve-Roperh et al., 2013; Rueda et al., 2002). In addition to neurons, it has been demonstrated that cannabinoids are modulators of astrogenesis and oligodendrogenesis in the early postnatal brain (Aguado et al., 2006; Gomez et al., 2011; Maccarrone et al., 2014) and that CB_1R activation promotes differentiation of astrocytes in cultured mouse neural progenitor cells (Aguado et al., 2006; Gómez Del Pulgar et al., 2002). The developmental ontogeny of endocannabinoids also parallels the pattern of cannabinoid receptor expression. Jiang et al. (2007) reported expression of CB_1Rs and CB_2Rs and associated ligands during hematopoietic differentiation of murine-derived embryoid bodies. When receptor expression was inhibited by specific antagonists, cell viability was significantly compromised, suggesting a role for receptor signaling in cell survival. These observations were supported by reports that CB_1R and CB_2R activation facilitated murine neural stem cell (NSC) and precursor cell proliferation in vitro (Compagnucci et al., 2013; Galve-Roperh et al., 2013; Molina-Holgado et al., 2007; Palazuelos et al., 2006; Trazzi et al., 2010). CB_1R expression levels are high in early life during neuronal differentiation and decrease with system maturation. In the human brain, CB_1R immunoreactivity has been detected in the cortical plate beginning as early as gestational week 9 then persisting throughout prenatal development. CB_1R is expressed in the subventricular zone and in Cajal-Retzius cells in the molecular zone of the developing neocortex (Zurolo et al., 2010). CB1R signaling serves to regulate radial and tangential migration of post-mitotic neurons to shape neuronal connectivity (Berghuis et al., 2005, 2007; Díaz-Alonso et al., 2017; Mulder et al., 2008; Saez et al., 2014). The basal activity of CB_1Rs plays a role in the initial development of dendrites and, indirectly, in axon initial segment development (Compagnucci et al., 2013; Tapia et al., 2017). Following successful migration, the CB_1R gradually localizes to developing axonal projections to facilitate synaptogenesis (Berghuis et al., 2007; Fernández-Ruiz et al., 2000; Mulder et al., 2008; Palazuelos et al., 2006; Zurolo et al., 2010).

This then triggers CB_1R internalization and elimination of the receptor from filopodia (Alpár et al., 2014; Berghuis et al., 2007). CB_1Rs modulate long-range axonal connectivity of pyramidal cells during corticogenesis by regulation of corticofugal axon navigation and fasciculation (Mulder et al., 2008; Watson et al., 2008; Wu et al., 2010). The value of balanced CBR signaling during development has been reported in several studies. Roland et al. (2014) reported that CB_1R activation by Δ^9-THC can rapidly transform the neuronal cytoskeleton by actomyosin contractility through a Rho-GTPase and Rho-associated kinase. A rapid neuronal remodeling was observed with retraction of neurites and axonal growth cones, elevated neuronal rigidity, and reshaping of somatodendritic morphology. Díaz-Alonso et al. (2017) reported regulation of pyramidal neuron migration by CB_1R/RhoA signaling. A deficiency in CB_1R signaling can result in malformations in cortical development mimicking subcortical band heterotopias (Díaz-Alonso et al., 2017). Reactive astrocytes in proximity to dysplastic neurons expressed CB_1Rs while microglia expressed CB_2Rs (Zurolo et al., 2010). Other investigators demonstrated that prenatal exposure to the synthetic cannabinoid, WIN 55,212-2 or Δ^9-THC induced permanent alterations in hippocampal long-term potentiation, hippocampal glutamate release, cortical glutamatergic neurotransmission, and long-lasting learning and memory impairments in adult offspring (Antonelli et al., 2005; Mereu et al., 2003). While neurons represent the primary developmental effects of CB_1R activation, there are data suggesting that developing astrocytes are also compromised by early exposure to Δ^9-THC (Suárez et al., 2002).

The available data on early exposure combined with knockout studies targeting enzymes involved in the biosynthesis and degradation of endocannabinoids (Aguado et al., 2005; Gao et al., 2010), CB_1R (Jin et al., 2004), and CB_2R (Palazuelos et al., 2006) indicate the endocannabinoid system as a key player in the processes underlying developmental influences on adult neurological function. In the developing mouse forebrain, DAGLs are located on elongating axons of projection neurons and are coexpressed with CB_1Rs, suggesting a role in axonal growth cone guidance (Bisogno et al., 2003). After the developmental process, DAGLs accumulate on postsynaptic dendrites and participate in endocannabinoid mediated modulation of synaptic strength (Keimpema et al., 2011). 2-AG shows equivalent levels in the developing and the adult brain (Berrendero et al., 1999). In comparison, AEA expression is low in the early prenatal brain and progressively increases during the perinatal period to higher levels seen in the adult brain.

Through CB_1R activation AEA plays a role in NSC differentiation into corticospinal motor neurons. The effects on brain function and behavior, mediated by cannabinoid signaling modulation during neurogenesis, are dependent on cannabinoid concentrations. Low concentrations of Δ^9-THC and AEA show no effect on neuronal and dopaminergic maturation, but AEA showed an enhancement of the frequency of synaptic activity. In contrast, higher doses of these CB_1R agonists reduced neuronal function by decreasing synaptic activity and ion currents (Galve-Roperh et al., 2007, 2013; Harkany et al., 2007; Stanslowsky et al., 2017; Zhou et al., 2014). Δ^9-THC exposure during pregnancy has been shown to interfere with corticospinal connectivity and to produce long-lasting alterations in the fine motor performance of the adult offspring and increased seizure susceptibility (de Salas-Quiroga et al., 2015). Maternal consumption of AEA resulted in subtle behavioral deficits in the adult offspring characterized by reduced cocaine-conditioned preference test, increased depressive-like behaviors, and impaired working memory (Amlani et al., 2017; Wu et al., 2014).

3.2 Subventricular zone and dentate gyrus neurogenesis

The ability to generate new neurons is maintained in discrete regions of the postnatal and mature brain including the subventricular zone (SVZ) of lateral ventricles, subgranular zone (SGZ) of the hippocampus (Bond et al., 2015; Ernst and Frisén, 2015). The low basal level of neurogenesis can be increased upon injury to the brain, i.e., injury-induced neurogenesis (Parent, 2003) or with actions such as exercise (Pereira et al., 2007). The process of generating new neurons involves cell proliferation, neuronal differentiation, and the integration of the new neurons into the neural networks. Neural stem cells (NSCs) of the SVZ and neural progenitor cells (NPCs) of the SGZ are highly sensitive to their microenvironment and extrinsic signaling molecules influence their proliferative activity and differentiation capacity. The successful integrating of new cells into a fully formed neural environment requires a tight level of regulation to prevent an excess of cells and maintain an optimal environment.

Endocannabinoids and receptor signaling can influence cell proliferation and overall neurogenesis in the mammalian brain (Kochman et al., 2006; Maćkowiak et al., 2007; Prenderville et al., 2015; Xapelli et al., 2013; Zimmermann et al., 2018). NPCs and NSCs express components of a functional endocannabinoid system to regulate neural progenitor proliferation

(Aguado et al., 2005; Butti et al., 2012; Compagnucci et al., 2013; Galve-Roperh et al., 2013). In the SVZ, ependymal and proliferating cells DAGL. In FAAH-deficient mice and loss of AEA, increased proliferation of SGZ NPCs was observed (Aguado et al., 2005); yet, neuroblast migration from the SVZ to the olfactory bulb was reduced with FAAH or DAGL inhibition (Goncalves et al., 2008). DAGLα regulates adult neurogenesis (Schuele et al., 2022) and inhibition of DAGL, eliminated cell proliferation in the SVZ (Goncalves et al., 2008) primarily due to loss of DAGLα (Gao et al., 2010). In comparison, in the SGZ, NPC proliferation was slightly diminished in DAGLα and DAGLβ knockout mouse lines (Gao et al., 2010). Thus, the various effects differ between neurogenic regions and cell type as well as proliferation vs migration.

Decreased NSC proliferation in the SGZ and SVZ has been demonstrated with genetic ablation of cannabinoid receptors or treatment with selective antagonists. In CB_2R deficient mice, the basal levels of SGZ NPC proliferation and generation of immature neurons were not altered as compared to wildtype mice (Mensching et al., 2019). However, reduced self-renewal capacity of cultured embryonic cortical NSCs was observed. Direct activation of cannabinoid receptors, including GPR55, stimulates the formation and maturation of new neurons in the SGZ and SVZ (Avraham et al., 2014; Goncalves et al., 2008; Hill et al., 2018; Jin et al., 2004; Palazuelos et al., 2006; Prenderville et al., 2015; Rivera et al., 2015; Rodrigues et al., 2017; Zimmermann et al., 2018). Further work showed a role for CB_1Rs in regulating the NSC pool and influencing dendritic morphology, activity-dependent plasticity, and behavior (Zimmermann et al., 2018). In an exercise model, treatment with the CB_1R antagonist, AM251, blocked hippocampal NPC proliferation (Hill et al., 2010). Increased differentiation of neurons, relative to glia by the CB_2R agonist, JWH133, was accompanied by an enhanced neurite outgrowth and inhibition of the ERK1/2 pathway (Compagnucci et al., 2013). Stimulation of CB_2Rs by the selective agonist, HU-308, increased proliferation of hippocampal NPC via phosphatidylinositol 3-kinase (PI3K)/Akt/mammalian target of rapamycin (mTOR)-dependent signaling (Palazuelos et al., 2012). This type of stimulation was also observed in glial fibrillary acidic protein (GFAP)/Gp120 mice with an attenuation of the genetic deficiency in neurogenesis observed with administration of the CB_2R agonist, AM1241 (Avraham et al., 2014). The synthetic cannabinoid, CBD has been shown to increase the proliferation, survival, differentiation, and migration of precursor cells in the SGZ (Campos et al., 2013; Esposito et al., 2011; Fogaça et al., 2018;

Luján et al., 2018; Schiavon et al., 2016). Numerous studies have found CBD increased both NSC proliferation and overall neurogenesis, with some indication that it may do so through interactions with CB_1Rs, CB_2Rs, and PPARγ (Campos et al., 2013; Esposito et al., 2011; Schiavon et al., 2016; Wolf et al., 2010). Rodrigues et al. (2017) demonstrated a critical role for interactions between CB_1Rs and CB_2Rs in cannabinoid-induced proliferation and differentiation of SVZ NSCs and SGZ NPCs. In a comparison study of THC and CBD, Wolf et al. (2010) reported that THC produced a learning deficit in mice but had no effect on adult neurogenesis. In CBD exposed mice, no effect was observed on learning, but there was an increase in neurogenesis. Esposito et al. (2011) also reported a stimulation of rat hippocampal neurogenesis with CBD possibly via interactions with the peroxisome proliferator-activated receptor alpha (PPARα).

CB_1Rs and CB_2Rs have been detected in oligodendrocyte progenitor cells (OPCs). 2-AG has been reported to enhance early OPC proliferation (Gomez et al., 2010, 2011; Molina-Holgado et al., 2002, 2007; Tomas-Roig et al., 2016). In oligodendrocytes, cannabinoid receptors affect progenitor survival and differentiation through phosphatidylinositol-3 kinase (PI3K)/Akt signaling (Arévalo-Martín et al., 2007; Molina-Holgado et al., 2002, 2007). The effect can be observed by activation of CB_1Rs or CB_2Rs using selective exogenous agonists which accelerate oligodendrocyte differentiation via the PI3K/Akt and mammalian target of rapamycin (mTOR) signaling pathways (Gomez et al., 2011, 2015). Pharmacological activation of CB_1Rs and CB_2Rs enhanced the expression of myelin basic protein by oligodendrocytes in subcortical white matter and promoted oligodendrogenesis in the SVZ of adult animals (Arévalo-Martín et al., 2007). The administration of WIN 55,212-2 was found to stimulate the survival of adult-born OPCs in animal models of viral encephalitis (Solbrig et al., 2010) and cuprizone demyelination (Tomas-Roig et al., 2016).

The overall data demonstrate that signaling via the endocannabinoid and cannabinoid receptors is a critical process in the normal development of the nervous system and contributes to the overall maintenance of the brain by providing a level of regulation of number and differentiation of cells from the neurogenic niches. The biological differences between cells generated during neurogenesis and those generated in the developed brain as well as differences in the source and type of cell present a challenge for interpretation of the exact pattern of effects from endocannabinoid signaling. In addition, the absence of extended studies to examine the fate of such cells limits the ability to determine the possible beneficial aspect of the responses. While proliferation of cells from each of these brain regions is often considered as a beneficial or neuroprotective process the random generation of new cells is not beneficial but rather requires modification and regulation to maintain an optimal balance of cell number and type. That aspect of the response to endocannabinoid system signaling remains to be explored.

4. Biological basis for potential adverse effects

The potential adverse effects of altered endocannabinoid system signaling in the nervous system are based within the critical role of the ECS to maintain balance whether in neuronal signaling or in microglia function. Based upon the current understanding of the mechanisms of action of cannabinoid receptor signaling one can consider possible pathways for adverse effects of prolonged exposure: (1) the inhibition of GABA and regulation of glutamate signaling, (2) influence on neural cell proliferation and differentiation, and (3) alterations in microglia responsivity. The more recent information demonstrating a role for ECS involvement with multiple neurotransmitter systems and the identification of additional ligand receptors opens a broader area for future research in the mechanisms of action of cannabinoid receptor activation. Gaining a better understanding of underlying processes affected will allow for an evaluation of the potential for beneficial as well as adverse effects of altering the normal balance.

4.1 Modifying GABAergic signaling

Activity of neuronal networks is maintained between complete silence and exceeding discharge by the interplay between synaptic excitation and inhibition. The excitatory feedback loops build up activity that subsequently require counterbalanced by dampening actions of the inhibitory feedback connections. The GABAergic interneuron is the workhorse for this ever-changing inhibition and is homeostatically regulated during physical and pathological alterations of network activity. In the neocortex, most GABA-containing neurons are local interneurons with few dendritic spines and classified as sparsely spiny, aspiny or, smooth cells within groups, e.g., basket cells, chandelier cells, double bouquet cells, local plexus neurons, or neurogliaform cells. These subtypes differ in morphology, neurochemical composition, somatic location, and terminal arborization.

GABA interneurons can also be classified by intrinsic membrane properties and synaptic connectivity. These heterogeneous neurons contribute to the organization of complex spatial and temporal patterns of network oscillations, selective input/output gating, suppression of background activity, and precise timing of action potentials (LoTurco and Kriegstein, 1991; Owens et al., 1999). Inhibition of inhibitory neurons results in a net excitation of downstream target neurons. The active effects of cannabinoids such as Δ^9-THC are likely associated with the role of CB_1Rs to inhibit the release of primarily GABA but also glutamate (Gerdeman and Lovinger, 2001; Kelley and Thayer, 2004; Robbe et al., 2001; Roloff and Thayer, 2009; Shen and Thayer, 1999; Straiker and Mackie, 2005; Wilson and Nicoll, 2002) and possibly inhibit uptake of GABA and other neurotransmitters (Banerjee et al., 1975; Coull et al., 1997; Maneuf et al., 1996). In humans, marijuana can impair memory through the disruption of information encoding and recall of newly acquired information (Abel, 1971; Ranganathan and D'Souza, 2006; Wilson et al., 1994). These observations are supported in animal models with a CB_1R mediated impairment of spatial working memory and disruption of non-spatial operant learning and memory (Hampson and Deadwyler, 1998; Heyser et al., 1993; Lichtman et al., 1995; Niyuhire et al., 2007; Puighermanal et al., 2009; Varvel and Lichtman, 2002; Wise et al., 2009). With acute administration, synthetic cannabinoids, endocannabinoids, and Δ^9-THC have been shown to block long-term potentiation in the CA1 region in hippocampal slice cultures (Misner and Sullivan, 1999; Nowicky et al., 1987; Stella et al., 1997). With a more prolonged in vivo exposure to Δ^9-THC hippocampal LTP is diminished under conditions of withdrawal (Fan et al., 2010; Hoffman et al., 2007). Synthetic cannabinoids acting on the CB_1Rs also present with adverse effects on the nervous system (Davidson et al., 2017).

While GABA is considered as the primary inhibitory neurotransmitter in the brain, the functional properties of GABA receptor signaling in the immature brain are different and often opposite those observed in the mature brain (Owens and Kriegstein, 2002; Represa and Ben-Ari, 2005). Acting as a developmental signal, GABA has been implicated in the regulation of key developmental steps including cell proliferation, cell death, migration, synaptogenesis, and neural circuit refinement (Dieni et al., 2013; Wang and Kriegstein, 2009). The dynamics and critical role for GABA in development is supported by the report that GABA-releasing synapses comprise close to 50% of the synapses in the early postnatal cortex, a number that declines to 15% in the adult (De Felipe et al., 1997). Components of the GABA signaling system appear very early in corticogenesis with GABAA-receptors expressed by precursor cells in the neocortical proliferative zone. GABA subunit expression shifts between proliferative and postmitotic zones. In the embryonic cortical plate containing almost exclusively postmitotic neurons, α3, β2/3 and γ3 are the predominant GABA subunits with a low expression of other α-subunits (α1, α2, and α5), the γ2-subunit, and the δ-subunit (Laurie et al., 1992; Ma and Barker, 1995; Poulter et al., 1992, 1993). At these early ages, it has been reported that neurogenic radial glia represents the GABA-responsive precursor cells. GABA-containing interneurons form functional synapses and are spontaneously active in the early neocortex, while afferent activation of these cells by excitatory feedback and feedforward pathways remains poorly developed. It is thought that this depolarization of neuronal stem cells is due to their high levels of [Cl−]i (Ge et al., 2006; Liu et al., 2005; LoTurco et al., 1995; Plotkin et al., 1997). While the exact mechanisms and impacts of GABA signaling in brain development are still a point of study, alterations in signaling during development have been reported to disrupt network activity in various brain regions (Dzhala et al., 2012; Gustorff et al., 2021).

A comparison of proliferative effects of GABA in the embryonic cortex showed a promotion of cell proliferation in the ventricular zone while inhibiting proliferation in the SVZ (Haydar et al., 2000). GABA can inhibit cell cycle progression and maturation of neural precursors in neurospheres and organotypic brain slices (Nguyen et al., 2003) and in the adult neurogenic SVZ (Liu et al., 2005) and SGZ (Catavero et al., 2018; Ge et al., 2006; Giachino et al., 2014). In the adult SVZ, immature neurons synthesize and release GABA (Bolteus and Bordey, 2004; Liu et al., 2005; Stewart et al., 2002), which in turn activates GABAA receptors in stem cells and immature neurons. Tonic GABAA activation depolarizes stem cells, transiently increases intracellular Ca^{2+}, and reduces the number of proliferative stem cells (Liu et al., 2005). Thus, GABAA serves in a regulatory capacity. Neuronal migration from the adult SVZ is also influenced by GABAergic signaling (Bolteus and Bordey, 2004; Ge et al., 2006; Liu et al., 2005; Nguyen et al., 2003). In the process of synaptogenesis many newborn neurons express GABAA receptors before glutamate receptors and young neurons first receive GABAergic inputs before forming glutamatergic synapses in the neocortex (Ben-Ari, 2006; Hennou et al., 2002; Owens et al., 1999; Tyzio et al., 1999). It is thought that these spontaneous GABA depolarizations provide the first excitatory drive necessary for activity-dependent synapse formation (Ben-Ari, 2006). The process of

GABA in regulating synaptogenesis and the balance between excitation and inhibition in a developing circuit is likely representative of a neurotransmitter that can excite neurons without directly producing excitotoxicity (Takesian and Hensch, 2013). Overall, the data suggested that GABA functions to regulate progenitor or stem cell proliferation, migration, and maturation and that inhibition of GABA receptors during this process can alter the normal pattern of neuronal development and maturation of brain structures as well as any regenerative capacity of adult neurogenesis. Understanding the basic mechanisms and outcomes associated with developmental expression and activation of the GABAergic system is critical for identifying and understanding the potential effects of cannabinoid receptor signaling.

4.2 Recruitment of brain immune cells

The function of a response of neuroimmune cells is to return the system to homeostatic balance and thus, any system dysregulation is a concern. The current understanding of the dynamic nature of microglia and their involvement in critical functions of the nervous system has significantly changed over the last decade, encompassing actions associated with an inflammatory response and those associated with non-inflammatory and phagocytic responses. The details of microglia functions are outside the scope of this chapter and are available in numerous recent reviews (Aramideh et al., 2021; Frost and Schafer, 2016; Harry, 2021; Michell-Robinson et al., 2015; Nosi et al., 2021; Woodburn et al., 2021). However, they help set a framework for evaluating the impact of immune regulation by cannabinoid signaling through microglia. Cell-cell contacts, signaling, and phagocytic properties of microglia play a significant role in all aspects of brain development, remodeling, and repair. In the healthy brain, microglia serve to maintain homeostasis in that they clear aberrant proteins such as amyloid beta and apolipoprotein E. They rapidly respond to minor vascular insults to repair the injury, assist in synaptic plasticity and remodeling, and facilitate signals for myelination and remyelination. Dysregulation in any of these activities can lead to unwanted or adverse outcomes. Examples of mechanisms by which the immune system can become dysregulated include cellular and molecular interactions (e.g., mitochondrial function, nuclear factor signaling, intracellular stress, telomeres degradation, cytokines/mitokines/chemokines, damage associated molecular pattern molecules, genetic/epigenetic factors, metabolic and general health status, lifestyle and environmental stress. These alterations may induce an accumulation of pro-inflammatory proteins and failure to clear aberrant proteins and cellular debris from the brain. An alteration in functional capability can influence the impact of an adaptive remodeling response or a severe destructive response. One specific example of dysregulation is the process of immunosenescence in which the immune cells show a diminished ability to respond properly to a stimulus. This has been primarily examined within the framework of age-related changes in the immune system known as inflammaging (Baban et al., 2021; Franceschi et al., 2018; Fulop et al., 2018; Pizza et al., 2011; van Beek et al., 2019). Immune cell remodeling serves as a mechanism to respond to changes and challenges presented to the system over the lifespan and is influenced by a series of primary and secondary factors. These factors play a central role in determining the type, magnitude, and direction of adaptation during senescence. The concept of neuroinflammaging can be applied to any prolonged sequence of events that alters the normal surveillance state of microglia (Gordleeva et al., 2020; Plaza-Zabala et al., 2017; Triviño and von Bernhardi, 2021). The long-lived nature of microglia represents a critical cell that can be modified by the repeated responses required of the cells over time leading to a compromised ability to function properly.

In general, the implication for cannabinoid signaling predominantly through CB_2Rs on microglia is one of downregulation of an inflammatory response and altered phagocytic activity (Aso et al., 2016; Ehrhart et al., 2005; Tolón et al., 2009; Young and Denovan-Wright, 2022). The purported anti-inflammatory actions of CB_2R activation serves as the basis for development of synthetic agonists as therapeutic agents. Given the immune regulating properties of specific phytocannabinoids such as CBD, they have been proposed as an effective adjunctive therapy to slow or reverse age-related pathologies addressing issues of neuroinflammaging (Baban et al., 2021). However, any short-term beneficial effects may be compromised or adverse effects manifest with prolonged downregulation of the inflammatory and/or phagocytic properties of microglia. Additional reports suggest that cannabinoids can serve as regulators of inflammasome signaling and thus address issues related to chronic inflammation (Suryavanshi et al., 2021). Inflammasomes are multiprotein complexes that involve the common adaptor, apoptosis-associated speck-like protein containing a CARD (ASC), and an effector, caspase 1. In the cytosol of immune cells, this complex forms in response to external signals and serves as a key sensor and effector of inflammation (Franchi et al., 2009;

Ghiringhelli et al., 2009; Gross et al., 2011). Activated inflammasomes induce caspase-1-dependent processing of the pro-inflammatory cytokines interleukin (IL)-1β and IL-18 (Broz and Dixit, 2016; Ghiringhelli et al., 2009), which play critical roles for host defense actions against invading factors and in coordinating the inflammatory response (Dinarello, 2009). This process allows for a priming of the system to provide a rapid full response as needed. However, it can also lead to an overactive process and contribute to chronic inflammation. While a therapeutic intervention of cannabinoids against neuroinflammation has been considered, targeting the downregulation of microglia or inflammasome activation requires a greater appreciation and understanding of the "friend or foe" aspect of these cells and their actions. The biological contributions of these cells across multiple aspects of brain development, function, and repair are critical and thus, any modification of this function for therapeutic use requires consideration of the roles of these cells.

5. Conclusions

The increased use of cannabinoid drugs for recreational or therapeutic purposes raises the concern for adverse effects with prolonged exposure, exposure during critical lifestages (e.g., developmental, adolescence, elderly), with comorbidities, or polypharmacy. The function of cannabionoids on the GABAergic system clearly raises a level of concern given what is known about the critical role of GABA in nervous system development and function. The more recent data demonstrating a functional role on additional neurotransmitter systems adds to such concern. The focus on therapeutic interventions designed to interact with the function of immune cells clearly requires additional thought and understanding of the role of these cells within the nervous system. Refraining from a simplistic approach and delving into the various aspects of the systems will offer success in identifying therapeutic strategies with minimal opportunity for adverse effects.

References

Abel, E.L., 1971. Retrieval of information after use of marihuana. Nature 231, 58.

Aguado, T., Monory, K., Palazuelos, J., et al., 2005. The endocannabinoid system drives neural progenitor proliferation. FASEB J. 19, 1704–1706.

Aguado, T., Palazuelos, J., Monory, K., et al., 2006. The endocannabinoid system promotes astroglial differentiation by acting on neural progenitor cells. J. Neurosci. 26 (5), 1551–1561.

Ahn, K.H., Mahmoud, M.M., Shim, J.Y., et al., 2013. Distinct roles of beta-arrestin 1 and beta-arrestin 2 in ORG27569-induced biased signaling and internalization of the cannabinoid receptor 1 (CB1). J. Biol. Chem. 288, 9790–9800.
Alger, B.E., 2012. Endocannabinoids at the synapse a decade after the dies mirabilis (29 march 2001) what we still do not know. J. Physiol. 590, 2203–2212.
Alpár, A., Tortoriello, G., Calvigioni, D., et al., 2014. Endocannabinoids modulate cortical development by configuring Slit2/Robo1 signalling. Nat. Commun. 5, 4421.
Aly, E., Masocha, W., 2021. Targeting the endocannabinoid system for management of HIV-associated neuropathic pain. A systematic review. IBRO Neurosci. Rep. 10, 109–118.
Amlani, A., Hornick, M.G., Cooper, K., et al., 2017. Maternal cannabinoid use alters cannabinoid (CB1) and endothelin (ETB) receptor expression in the brains of dams but not their offspring. Dev. Neurosci. 39, 498–506.
Antonelli, T., Tomasini, M.C., Tattoli, M., et al., 2005. Prenatal exposure to the CB1 receptor agonist WIN 55,212-2 causes learning disruption associated with impaired cortical NMDA receptor function and emotional reactivity changes in rat offspring. Cereb. Cortex 15, 2013–2020.
Aramideh, J.A., Vidal-Itriago, A., Morsch, M., et al., 2021. Cytokine Signalling at the microglial Penta-partite synapse. Int. J. Mol. Sci. 22 (24), 13186.
Araque, A., Castillo, P.E., Manzoni, O.J., et al., 2017. Synaptic functions of endocannabinoid signaling in health and disease. Neuropharmacology 124, 13–24.
Arévalo-Martín, Á., García-Ovejero, D., Rubio-Araiz, A., et al., 2007. Cannabinoids modulate Olig2 and polysialylated neural cell adhesion molecule expression in the subventricular zone of post-natal rats through cannabinoid receptor 1 and cannabinoid receptor 2. Eur. J. Neurosci. 26, 1548–1559.
Ashton, C.H., 2001. Pharmacology and effects of cannabis, a brief review. Br. J. Psychiatry 178, 101–106.
Asimaki, O., Mangoura, D., 2011. Cannabinoid receptor 1 induces a biphasic ERK activation via multiprotein signaling complex formation of proximal kinases PKCepsilon, Src, and Fyn in primary neurons. Neurochem. Int. 58, 135–144.
Aso, E., Juvés, S., Maldonado, R., et al., 2013. CB2 cannabinoid receptor agonist ameliorates Alzheimer-like phenotype in AβPP/PS1 mice. J. Alzheimers Dis. 35, 847–858.
Aso, E., Andrés-Benito, P., Carmona, M., et al., 2016. Cannabinoid receptor 2 participates in amyloid-β processing in a mouse model of Alzheimer's disease but plays a minor role in the therapeutic properties of a cannabis-based medicine. J. Alzheimers Dis. 51, 489–500.
Atwood, B.K., Mackie, K., 2010. CB2, a cannabinoid receptor with an identity crisis. Br. J. Pharmacol. 160 (3), 467–479.
Aung, M.M., Griffin, G., Huffman, J.W., et al., 2000. Influence of the N-1 alkyl chain length of cannabimimetic indoles upon CB(1) and CB(2) receptor binding. Drug Alcohol Depend. 60 (2), 133–140.
Avraham, H.K., Jiang, S., Fu, Y., et al., 2014. The cannabinoid CB_2 receptor agonist AM1241 enhances neurogenesis in GFAP/Gp120 transgenic mice displaying deficits in neurogenesis. Br. J. Pharmacol. 171 (2), 468–479.
Baban, B., Khodadadi, H., Salles, É.L., et al., 2021. Inflammaging and Cannabinoids. Ageing Res. Rev. 2021 (72), 101487.
Bacci, A., Huguenard, J.R., Prince, D.A., 2004. Long-lasting self-inhibition of neocortical interneurons mediated by endocannabinoids. Nature 431, 312–316.
Bagher, A.M., Young, A.P., Laprairie, R.B., 2020. Heteromer formation between cannabinoid type 1 and dopamine type 2 receptors is altered by combination cannabinoid and antipsychotic treatments. J. Neurosci. Res. 98, 2496–2509.
Banerjee, S.P., Snyder, S.H., Mechoulam, R., 1975. Cannabinoids, influence on neurotransmitter uptake in rat brain synaptosomes. J. Pharmacol. Exp. Ther. 194, 74–81.

Bara, A., Ferland, J.M.N., Rompala, G., et al., 2021. Cannabis and synaptic reprogramming of the developing brain. Nat. Rev. Neurosci. 22, 423–438.
Bari, M., Battista, N., Fezza, F., et al., 2005. Lipid rafts control signaling of type-1 cannabinoid receptors in neuronal cells. Implications for anandamide-induced apoptosis. J. Biol. Chem. 280 (13), 12212–12220.
Barnett-Norris, J., Lynch, D., Reggio, P.H., 2005. Lipids, lipid rafts and caveolae, their importance for GPCR signaling and their centrality to the endocannabinoid system. Life Sci. 77 (14), 1625–1639.
Batkai, S., Jarai, Z., Wagner, J.A., et al., 2001. Endocannabinoids acting at vascular CB1 receptors mediate the vasodilated state in advanced liver cirrhosis. Nat. Med. 7, 827–832.
Beltramo, M., Stella, N., Calignano, A., et al., 1997. Functional role of high-affinity anandamide transport, as revealed by selective inhibition. Science 277, 1094–1097.
Benard, G., Massa, F., Puente, N., et al., 2012. Mitochondrial CB(1) receptors regulate neuronal energy metabolism. Nat. Neurosci. 15, 558–564.
Ben-Ari, Y., 2006. Basic developmental rules and their implications for epilepsy in the immature brain. Epileptic Disord. 8, 91–102.
Benito, C., Núñez, E., Tolón, R.M., et al., 2003. Cannabinoid CB2 receptors and fatty acid amide hydrolase are selectively overexpressed in neuritic plaque-associated glia in Alzheimer's disease brains. J. Neurosci. 23, 11136–11141.
Berghuis, P., Dobszay, M.B., Wang, X., et al., 2005. Endocannabinoids regulate interneuron migration and morphogenesis by transactivating the TrkB receptor. Proc. Natl. Acad. Sci. U. S. A. 102, 19115–19120.
Berghuis, P., Rajnicek, A.M., Morozov, Y.M., et al., 2007. Hardwiring the brain. Endocannabinoids shape neuronal connectivity. Science 316, 1212–1216.
Berrendero, F., Sepe, N., Ramos, J.A., et al., 1999. Analysis of cannabinoid receptor binding and mRNA expression and endogenous cannabinoid contents in the developing rat brain during late gestation and early postnatal period. Synapse 33, 181–191.
Bisogno, T., Howell, F., Williams, G., et al., 2003. Cloning of the first sn1-DAG lipases points to the spatial and temporal regulation of endocannabinoid signaling in the brain. J. Cell Biol. 163, 463–468.
Blankman, J.L., Simon, G.M., Cravatt, B.F., 2007. A comprehensive profile of brain enzymes that hydrolyze the endocannabinoid 2-arachidonoylglycerol. Chem. Biol. 14, 1347–1356.
Blankman, J.L., Long, J.Z., Trauger, S.A., et al., 2013. ABHD12 controls brain lysophosphatidylserine pathways that are deregulated in a murine model of the neurodegenerative disease PHARC. Proc. Natl. Acad. Sci. U. S. A. 110, 1500–1505.
Blume, L.C., Leone-Kabler, S., Luessen, D.J., et al., 2016. Cannabinoid receptor interacting protein suppresses agonist-driven CB1 receptor internalization and regulates receptor replenishment in an agonist-biased manner. J. Neurochem. 139, 396–407.
Bodor, A.L., Katona, I., Nyiri, G., et al., 2005. Endocannabinoid signaling in rat somatosensory cortex, laminar differences and involvement of specific interneuron types. J. Neurosci. 25, 6845–6856.
Bolteus, A.J., Bordey, A., 2004. GABA release and uptake regulate neuronal precursor migration in the postnatal subventricular zone. J. Neurosci. 24, 7623–7631.
Bond, A.M., Ming, G.L., Song, H., 2015. Adult mammalian neural stem cells and neurogenesis, five decades later. Cell Stem Cell 17, 385–395.
Bourque, J., Potvin, S., 2021. Cannabis and cognitive functioning, from acute to residual effects, from randomized controlled trials to prospective designs. Front. Psych. 12, 596601.
Breivogel, C.S., Selley, D.E., Childers, S.R., 1998. Cannabinoid receptor agonist efficacy for stimulating [35S]GTPγS binding to rat cerebellar membranes correlates with agonist-induced decreases in GDP affinity. J. Biol. Chem. 273, 16865–16873.

Breivogel, C.S., Puri, V., Lambert, J.M., et al., 2013. The influence of beta-arrestin2 on cannabinoid CB1 receptor coupling to G-proteins and subcellular localization and relative levels of beta-arrestin1 and 2 in mouse brain. J. Recept. Signal Transduct. Res. 33, 367–379.

Brents, L.K., Gallus-Zawada, A., Radominska-Pandya, A., 2012. Monohydroxylated metabolites of the K2 synthetic cannabinoid JWH-073 retain intermediate to high cannabinoid 1 receptor (CB1R) affinity and exhibit neutral antagonist to partial agonist activity. Biochem. Pharmacol. 83 (7), 952–961.

Britch, S.C., Babalonis, S., Walsh, S.L., 2020. Cannabidiol, pharmacology and therapeutic targets. Psychopharmacologia 238, 9–28.

Brown, A.J., 2007. Novel cannabinoid receptors. Br. J. Pharmacol. 152, 567–575.

Broz, P., Dixit, V.M., 2016. Inflammasomes, mechanism of assembly, regulation and signalling. Nat. Rev. Immunol. 16 (7), 407–420.

Brucki, S., Adoni, T., Almeida, C., et al., 2021. Cannabinoids in neurology - position paper from scientific departments from Brazilian academy of neurology. Arq. Neuropsiquiatr. 79 (4), 354–369.

Buckley, N.E., Hansson, S., Harta, G., et al., 1998. Expression of the CB1 and CB2 receptor messenger RNAs during embryonic development in the rat. Neuroscience 82, 1131–1149.

Buckley, N.E., McCoy, K.L., Mezey, E., et al., 2000. Immunomodulation by cannabinoids is absent in mice deficient for the cannabinoid CB(2) receptor. Eur. J. Pharmacol. 396 (2–3), 141–149.

Butti, E., Bacigaluppi, M., Rossi, S., et al., 2012. Subventricular zone neural progenitors protect striatal neurons from glutamatergic excitotoxicity. Brain 135, 3320–3335.

Cabanero, D., Martin-Garcia, E., Maldonado, R., 2021. The CB2 cannabinoid receptor as a therapeutic target in the central nervous system. Expert Opin. Ther. Targets 25, 659–676.

Cabañero, D., Ramírez-López, A., Drews, E., et al., 2020. Protective role of neuronal and lymphoid cannabinoid CB2 receptors in neuropathic pain. Elife 9, e55582.

Cadas, H., Gaillet, S., Beltramo, M., et al., 1996. Biosynthesis of an endogenous cannabinoid precursor in neurons and its control by calcium and cAMP. J. Neurosci. 16 (12), 3934–3942.

Calandra, B., Portier, M., Kernéis, A., et al., 1999. Dual intracellular signaling pathways mediated by the human cannabinoid CB1 receptor. Eur. J. Pharmacol. 374 (3), 445–455.

Callén, L., Moreno, E., Barroso-Chinea, P., et al., 2012. Cannabinoid receptors CB1 and CB2 form functional heteromers in brain. J. Biol. Chem. 287, 20851–20865.

Campos, A.C., Ortega, Z., Palazuelos, J., et al., 2013. The anxiolytic effect of cannabidiol on chronically stressed mice depends on hippocampal neurogenesis, involvement of the endocannabinoid system. Int. J. Neuropsychopharmacol. 16, 1407–1419.

Carlisle, S.J., Cabral, F.M., Staab, A., et al., 2002. Differential expression of the CB2 cannabinoid receptor by rodent macrophages and macrophage-like cells in relation to cell activation. Int. Immunopharmacol. 2, 69–82.

Castillo, P.E., Younts, T.J., Chávez, A.E., et al., 2012. Endocannabinoid signaling and synaptic function. Neuron 76 (1), 70–81.

Catavero, C., Bao, H., Song, J., 2018. Neural mechanisms underlying GABAergic regulation of adult hippocampal neurogenesis. Cell Tissue Res. 371, 33–46.

Chávez, A.E., Chiu, C.Q., Castillo, P.E., 2010. TRPV1 activation by endogenous anandamide triggers postsynaptic long-term depression in dentate gyrus. Nat. Neurosci. 13 (12), 1511–1518.

Chen, J.K., Chen, J., Imig, J.D., et al., 2008. Identification of novel endogenous cytochrome p450 arachidonate metabolites with high affinity for cannabinoid receptors. J. Biol. Chem. 283 (36), 24514–24524.

Chen, R., Zhang, J., Fan, N., et al., 2013. Delta 9-THC-caused synaptic and memory impairments are mediated through COX-2 signaling. Cell 155, 1154–1165.

Chen, X., Zheng, C., Qian, J., et al., 2014. Involvement of beta-arrestin-2 and clathrin in agonist-mediated internalization of the human cannabinoid CB2 receptor. Curr. Mol. Pharmacol. 7, 67–80.

Chen, D.J., Gao, M., Gao, F.F., et al., 2017. Brain cannabinoid receptor 2, expression, function and modulation. Acta Pharmacol. Sin. 38, 312–316.

Chevaleyre, V., Castillo, P.E., 2004. Endocannabinoid-mediated metaplasticity in the hippocampus. Neuron 43, 871–881.

Compagnucci, C., Di Siena, S., Bustamante, M.B., et al., 2013. Type-1 (CB1) cannabinoid receptor promotes neuronal differentiation and maturation of neural stem cells. PLoS One 8, e54271.

Cong, J., Lu, K., Zou, W., et al., 2021. Astroglial CB1 cannabinoid receptors mediate CP 55,940-induced conditioned place aversion through cyclooxygenase-2 signaling in mice. Front. Cell. Neurosci. 15, 772549.

Constantinescu, C.S., Farooqi, N., O'Brien, K., et al., 2011. Experimental autoimmune encephalomyelitis (EAE) as a model for multiple sclerosis (MS). Br. J. Pharmacol. 164 (4), 1079–1106.

Correa, F., Hernangómez, M., Mestre, L., et al., 2010. Anandamide enhances IL-10 production in activated microglia by targeting CB(2) receptors: roles of ERK1/2, JNK, and NF-kappaB. Glia 58 (2), 135–147.

Coull, M.A., Johnston, A.T., Pertwee, R.G., et al., 1997. Action of Delta-9-tetrahydrocannabinol on GABAA receptor- mediated responses in a grease-gap recording preparation of the rat hippocampal slice. Neuropharmacology 36, 1387–1392.

Cox, M.L., Haller, V.L., Welch, S.P., 2007. The antinociceptive effect of Delta9-tetrahydrocannabinol in the arthritic rat involves the CB(2) cannabinoid receptor. Eur. J. Pharmacol. 570, 50–56.

Cravatt, B.F., Giang, D.K., Mayfield, S.P., 1996. Molecular characterization of an enzyme that degrades neuromodulatory fatty-acid amides. Nature 384, 83–87.

Daigle, T.L., Kwok, M.L., Mackie, K., 2008. Regulation of CB1 cannabinoid receptor internalization by a promiscuous phosphorylation-dependent mechanism. J. Neurochem. 106 (1), 70–82.

Dalton, G.D., Howlett, A.C., 2012. Cannabinoid CB1 receptors transactivate multiple receptor tyrosine kinases and regulate serine/threonine kinases to activate ERK in neuronal cells. Br. J. Pharmacol. 165 (8), 2497–2511.

Davidson, C., Opacka-Juffry, J., Arevalo-Martin, A., et al., 2017. Spicing up pharmacology. A review of synthetic cannabinoids from structure to adverse events. Adv. Pharmacol. 80, 135–168.

Davis, M.I., Ronesi, J., Lovinger, D.M., 2003. A predominant role for inhibition of the adenylate cyclase/protein kinase a pathway in ERK activation by cannabinoid receptor 1 in N1E-115 neuroblastoma cells. J. Biol. Chem. 278 (49), 48973–48980.

De Felipe, J., Marco, P., Fairen, A., et al., 1997. Inhibitory synaptogenesis in mouse somatosensory cortex. Cereb. Cortex 7, 619–634.

de Salas-Quiroga, A., Díaz-Alonso, J., García-Rincón, D., et al., 2015. Prenatal exposure to cannabinoids evokes long-lasting functional alterations by targeting CB1 receptors on developing cortical neurons. Proc. Natl. Acad. Sci. U. S. A. 112, 13693–13698.

Di Marzo, V., 2008. Endocannabinoids, synthesis and degradation. Rev. Physiol. Biochem. Pharmacol. 160, 1–24.

Di Marzo, V., 2018. New approaches and challenges to targeting the endocannabinoid system. Nat. Rev. Drug Discov. 17 (9), 623–639.

Di Marzo, V., Breivogel, C.S., Tao, Q., et al., 2000. Levels, metabolism, and pharmacological activity of anandamide in CB(1) cannabinoid receptor knockout mice, evidence for non-CB(1), non-CB(2) receptor-mediated actions of anandamide in mouse brain. J. Neurochem. 75, 2434–2444.

Díaz-Alonso, J., Aguado, T., Wu, C.-S., et al., 2012a. The CB1 cannabinoid receptor drives corticospinal motor neuron differentiation through the Ctip2/Satb2 transcriptional regulation axis. J. Neurosci. 32, 16651–16665.

Díaz-Alonso, J., Guzmán, M., Galve-Roperh, I., 2012b. Endocannabinoids via CB_1 receptors act as neurogenic niche cues during cortical development. Philos. Trans. R. Soc. Lond. B Biol. Sci. 367, 3229–3241.

Díaz-Alonso, J., de Salas-Quiroga, A., Paraíso-Luna, J., et al., 2017. Loss of cannabinoid CB1 receptors induces cortical migration malformations and increases seizure susceptibility. Cereb. Cortex 27, 5303–5317.

Dieni, C.V., Chancey, J.H., Overstreet-Wadiche, L.S., 2013. Dynamic functions of GABA signaling during granule cell maturation. Front. Neural Circuits 6, 113.

Dinarello, C.A., 2009. Immunological and inflammatory functions of the interleukin-1 family. Annu. Rev. Immunol. 27, 519–550.

Dinh, T.P., Carpenter, D., Leslie, F.M., et al., 2002. Brain monoglyceride lipase participating in endocannabinoid inactivation. Proc. Natl. Acad. Sci. U. S. A. 99, 10819–10824.

Djeungoue-Petga, M.A., Hebert-Chatelain, E., 2017. Linking mitochondria and synaptic transmission. The CB1 receptor. Bioessays 39 (12).

Doğan, S., Akman, C., Yılmaz, A., et al., 2016. A hypothermic case with giant Osborn waves and atrial fibrillation after using synthetic cannabinoid. Anatol. J. Cardiol. 16 (2), E1.

Donvito, G., Nass, S.R., Wilkerson, J.L., et al., 2018. The endogenous cannabinoid system, a budding source of targets for treating inflammatory and neuropathic pain. Neuropsychopharmacology 43, 52–79.

Dos Santos, R.G., Hallak, J.E.C., Crippa, J.A.S., 2021. Neuropharmacological effects of the main phytocannabinoids, a narrative review. Adv. Exp. Med. Biol. 1264, 29–45.

Dowie, M.J., Grimsey, N.L., Hoffman, T., et al., 2014. Cannabinoid receptor CB2 is expressed on vascular cells, but not astroglial cells in the post-mortem human Huntington's disease brain. J. Chem. Neuroanat. 59-60, 62–71.

Duffy, C.M., Xu, H., Nixon, J.P., et al., 2017. Identification of a fatty acid binding protein4-UCP2 axis regulating microglial mediated neuroinflammation. Mol. Cell. Neurosci. 80, 52–57.

Dzhala, V., Valeeva, G., Glykys, J., et al., 2012. Traumatic alterations in GABA signaling disrupt hippocampal network activity in the developing brain. J. Neurosci. 32 (12), 4017–4031.

Egertová, M., Cravatt, B.F., Elphick, M.R., 2003. Comparative analysis of fatty acid amide hydrolase and CB1 cannabinoid receptor expression in the mouse brain, evidence of a widespread role for fatty acid amide hydrolase in regulation of endocannabinoid signaling. Neuroscience 119, 481–496.

Ehrhart, J., Obregon, D., Mori, T., et al., 2005. Stimulation of cannabinoid receptor 2 (CB2) suppresses microglial activation. J. Neuroinflammation 2, 29.

Ernst, A., Frisén, J., 2015. Adult neurogenesis in humans - common and unique traits in mammals. PLoS Biol. 2015 (13), 1–12.

Esposito, G., Scuderi, C., Valenza, M., et al., 2011. Cannabidiol reduces Aβ-induced neuroinflammation and promotes hippocampal neurogenesis through PPARγ involvement. PLoS One 6 (12), e28668.

Fan, N., Yang, H., Zhang, J., et al., 2010. Reduced expression of glutamate receptors and phosphorylation of CREB are responsible for in vivo Delta9-THC exposure-impaired hippocampal synaptic plasticity. J. Neurochem. 112, 691–702.

Fegley, D., Kathuria, S., Mercier, R., et al., 2004. Anandamide transport is independent of fatty-acid amide hydrolase activity and is blocked by the hydrolysis-resistant inhibitor AM1172. Proc. Natl. Acad. Sci. U. S. A. 101, 8756–8761.
Felder, C.C., Glass, M., 1998. Cannabinoid receptors and their endogenous agonists. Annu. Rev. Pharmacol. Toxicol. 38, 179–200.
Ferguson, S.S., 2001. Evolving concepts in G protein-coupled receptor endocytosis, the role in receptor desensitization and signaling. Pharmacol. Rev. 53, 1–24.
Fernández-Ruiz, J., Berrendero, F., Hernández, M.L., et al., 2000. The endogenous cannabinoid system and brain development. Trends Neurosci. 23, 14–20.
Ferrisi, R., Ceni, C., Bertini, S., et al., 2021. Medicinal chemistry approach, pharmacology and neuroprotective benefits of CB2R modulators in neurodegenerative diseases. Pharmacol. Res. 170, 105607.
Fields, J., Swinton, M., Montilla-Perez, P., et al., 2022. The cannabinoid receptor agonist, WIN-55212-2, suppresses the activation of proinflammatory genes induced by interleukin 1 beta in human astrocytes. Cannabis Cannabinoid Res. 7, 78–92. https://doi.org/10.1089/can.2020.0128.
Fiskerstrand, T., H'Mida-Ben Brahim, D., Johansson, S., et al., 2010. Mutations in ABHD12 cause the neurodegenerative disease PHARC. An inborn error of endocannabinoid metabolism. Am. J. Hum. Genet. 87, 410–417.
Fletcher-Jones, A., Hildick, K.L., Evans, A.J., et al., 2020. Protein interactors and trafficking pathways that regulate the cannabinoid type 1 receptor (CB1R). Front. Mol. Mol. Neurosci. 13, 108.
Foerster, S., Guzman de la Fuente, A., Kagawa, Y., et al., 2020. The fatty acid binding protein FABP7 is required for optimal oligodendrocyte differentiation during myelination but not during remyelination. Glia 68, 1410–1420.
Fogaça, M.V., Campos, A.C., Coelho, L.D., et al., 2018. The anxiolytic effects of cannabidiol in chronically stressed mice are mediated by the endocannabinoid system. Role of neurogenesis and dendritic remodeling. Neuropharmacology 135, 22–33.
Ford, B.M., Tai, S., Fantegrossi, W.E., 2017. Synthetic pot. Not your grandfather's marijuana. Trends Pharmacol. Sci. 38, 257–276.
Franceschi, C., Garagnani, P., Parini, P., et al., 2018. Inflammaging, a new immune–metabolic viewpoint for age-related diseases. Nat. Rev. Endocrinol. 14, 576–590.
Franchi, L., Eigenbrod, T., Muñoz-Planillo, R., et al., 2009. The inflammasome, a caspase-1-activation platform that regulates immune responses and disease pathogenesis. Nat. Immunol. 10 (3), 241–247.
Franco, R., Reyes-Resina, I., Aguinaga, D., et al., 2019. Potentiation of cannabinoid signaling in microglia by adenosine A 2A receptor antagonists. Glia 67, 2410–2423.
Freund, T.F., Katona, I., Piomelli, D., 2003. Role of endogenous cannabinoids in synaptic signaling. Physiol. Rev. 83 (3), 1017–1066.
Frost, J.L., Schafer, D.P., 2016. Microglia, architects of the developing nervous system. Trends Cell Biol. 26, 587–597.
Fulop, T., Larbi, A., Dupuis, G., et al., 2018. Immunosenescence and Inflamm-aging as two sides of the same coin. Friends or foes? Front. Immunol. 8, 1960.
Gabrielli, M., Battista, N., Riganti, L., et al., 2015. Active endocannabinoids are secreted on extracellular membrane vesicles. EMBO Rep. 16, 213–220.
Gaffuri, A.L., Ladarre, D., Lenkei, Z., 2012. Type-1 cannabinoid receptor signaling in neuronal development. Pharmacology 90 (1–2), 19–39.
Galiègue, S., Mary, S., Marchand, J., et al., 1995. Expression of central and peripheral cannabinoid receptors in human immune tissues and leukocyte subpopulations. Eur. J. Biochem. 232, 54–61.

Galve-Roperh, I., Aguado, T., Palazuelos, J., et al., 2007. The endocannabinoid system and neurogenesis in health and disease. Neuroscientist 13, 109–114.
Galve-Roperh, I., Chiurchiù, V., Díaz-Alonso, J., et al., 2013. Cannabinoid receptor signaling in progenitor/stem cell proliferation and differentiation. Prog. Lipid Res. 52 (4), 633–650.
Gamage, T.F., Farquhar, C.E., Lefever, T.W., et al., 2018. Molecular and behavioral pharmacological characterization of abused synthetic cannabinoids MMB- and MDMB-FUBINACA, MN-18, NNEI, CUMYL-PICA, and 5-Fluoro-CUMYL-PICA. J. Pharmacol. Exper Therap. 365 (2), 437–446.
Gao, Y., Vasilyev, D.V., Goncalves, M.B., et al., 2010. Loss of retrograde endocannabinoid signaling and reduced adult neurogenesis in diacylglycerol lipase knock-out mice. J. Neurosci. 30, 2017–2024.
Garcia, A.B., Soria-Gomez, E., Bellocchio, L., et al., 2016. Cannabinoid receptor type-1, breaking the dogmas. F1000 Fac Rev. 5, 990.
Ge, S., Goh, E.L., Sailor, K.A., et al., 2006. GABA regulates synaptic integration of newly generated neurons in the adult brain. Nature 439, 589–593.
Gerdeman, G., Lovinger, D.M., 2001. CB1 cannabinoid receptor inhibits synaptic release of glutamate in rat dorsolateral striatum. J. Neurophysiol. 85, 468–471.
Gerostamoulos, D., Drummer, O.H., Woodford, N.W., 2015. Deaths linked to synthetic cannabinoids. Forensic Sci. Med. Pathol. 11, 478.
Gerstner, J.R., Perron, I.J., Riedy, S.M., et al., 2017. Normal sleep requires the astrocyte brain-type fatty acid binding protein FABP7. Sci. Adv. 3, e1602663.
Ghiringhelli, F., Apetoh, L., Tesniere, A., et al., 2009. Activation of the NLRP3 inflammasome in dendritic cells induces IL-1beta-dependent adaptive immunity against tumors. Nat. Med. 15 (10), 1170–1178.
Giachino, C., Barz, M., Tchorz, J.S., et al., 2014. GABA suppresses neurogenesis in the adult hippocampus through GABAB receptors. Development 141, 83–90.
Giacoppo, S., Pollastro, F., Grassi, G., et al., 2017. Target regulation of PI3K/Akt/mTOR pathway by cannabidiol in treatment of experimental multiple sclerosis. Fitoterapia 116, 77–84.
Giuffrida, A., Parsons, L.H., Kerr, T.M., et al., 1999. Dopamine activation of endogenous cannabinoid signaling in dorsal striatum. Nat. Neurosci. 2, 358–363.
Giuffrida, A., Beltramo, M., Piomelli, D., 2001. Mechanisms of endocannabinoid inactivation, biochemistry and pharmacology. J. Pharmacol. Exper. Therap. 298, 7–14.
Gómez Del Pulgar, T., De Ceballos, M.L., Guzmán, M., et al., 2002. Cannabinoids protect astrocytes from ceramide-induced apoptosis through the phosphatidylinositol 3-kinase/protein kinase B pathway. J. Biol. Chem. 277 (39), 36527–36533.
Gomez, O., Arevalo-Martin, A., Garcia-Ovejero, D., et al., 2010. The constitutive production of the endocannabinoid 2-arachidonoylglycerol participates in oligodendrocyte differentiation. Glia 58 (16), 1913–1927.
Gomez, O., Sanchez-Rodriguez, A., Le, M., et al., 2011. Cannabinoid receptor agonists modulate oligodendrocyte differentiation by activating PI3K/Akt and the mammalian target of rapamycin (mTOR) pathways. Br. J. Pharmacol. 163 (7), 1520–1532.
Gomez, O., Sanchez-Rodriguez, M.A., Ortega-Gutierrez, S., et al., 2015. A basal tone of 2-arachidonoylglycerol contributes to early oligodendrocyte progenitor proliferation by activating phosphatidylinositol 3-kinase (PI3K)/AKT and the mammalian target of rapamycin (MTOR) pathways. J. Neuroimmune Pharmacol. 10 (2), 309–317.
Gonca, E., Darıcı, F., 2015. The effect of cannabidiol on ischemia/reperfusion-induced ventricular arrhythmias. The role of adenosine A1 receptors. J. Cardiovasc. Pharmacol. Ther. 20, 76–83.

Goncalves, M.B., Suetterlin, P., Yip, P., et al., 2008. A diacylglycerol lipase-CB2 cannabinoid pathway regulates adult subventricular zone neurogenesis in an age-dependent manner. Mol. Cell. Neurosci. 38, 526–536.
Gong, J.-P., Onaivi, E.S., Ishiguro, H., Liu, Q.-R., et al., 2006. Cannabinoid CB2 receptors, immunohistochemical localization in rat brain. Brain Res. 1071, 10–23.
Gordleeva, S., Kanakov, O., Ivanchenko, M., et al., 2020. Brain aging and garbage cleaning. Modelling the role of sleep, glymphatic system, and microglia senescence in the propagation of inflammaging. Semin. Immunopathol. 42 (5), 647–665.
Grim, T.W., Morales, A.J., Gonek, M.M., et al., 2016. Stratification of cannabinoid 1 receptor (CB1R) agonist efficacy, manipulation of CB1R density through use of transgenic mice reveals congruence between in vivo and in vitro assays. J. Pharmacol. Exp. Ther. 359, 329–339.
Grimsey, N.L., Graham, E.S., Dragunow, M., et al., 2010. Cannabinoid receptor 1 trafficking and the role of the intracellular pool, implications for therapeutics. Biochem. Pharmacol. 80, 1050–1062.
Gross, O., Thomas, C.J., Guarda, G., et al., 2011. The inflammasome, an integrated view. Immunol. Rev. 243 (1), 136–151.
Guindon, J., Hohmann, A.G., 2009. The endocannabinoid system and pain. CNS Neurol. Disord. Drug Targets 8 (6), 403–421.
Gulyas, A.I., Cravatt, B.F., Bracey, M.H., Dinh, T.P., et al., 2004. Segregation of two endocannabinoid-hydrolyzing enzymes into pre- and postsynaptic compartments in the rat hippocampus, cerebellum and amygdala. Eur. J. Neurosci. 20, 441–458.
Gustorff, C., Scheuer, T., Schmitz, T., et al., 2021. GABAB receptor-mediated impairment of intermediate progenitor maturation during postnatal hippocampal neurogenesis of newborn rats. Front. Cell. Neurosci. 15, 651072.
Hampson, R.E., Deadwyler, S.A., 1998. Role of cannabinoid receptors in memory storage. Neurobiol. Dis. 5, 474–482.
Han, J., Kesner, P., Metna-Laurent, M., et al., 2012. Acute cannabinoids impair working memory through astroglial CB1 receptor modulation of hippocampal LTD. Cell 48, 1039–1050.
Han, S., Chen, J.J., Chen, J.Z., 2014. Latest progress in the identification of novel synthetic ligands for the cannabinoid CB2 receptor. Mini Rev. Med. Chem. 14 (5), 426–443.
Harkany, T., Dobszay, M.B., Cayetanot, F., et al., 2005. Redistribution of CB1 cannabinoid receptors during evolution of cholinergic basal forebrain territories and their cortical projection areas, a comparison between the gray mouse lemur (Microcebus murinus, primates) and rat. Neuroscience 135 (2), 595–609.
Harkany, T., Guzmán, M., Galve-Roperh, I., et al., 2007. The emerging functions of endocannabinoid signaling during CNS development. Trends Pharmacol. Sci. 28, 83–92.
Harry, G.J., 2021. Microglia in neurodegenerative events—an initiator or a significant other? Int. J. Mol. Sci. 22 (11), 5818.
Hashiesh, H.M., Sharma, C., Goyal, S.N., et al., 2021. A focused review on CB2 receptor-selective pharmacological properties and therapeutic potential of β-caryophyllene, a dietary cannabinoid. Biomed. Pharmacotherapy Biomed. Pharmacother. 140, 111639.
Haydar, T.F., Wang, F., Schwartz, M.L., et al., 2000. Differential modulation of proliferation in the neocortical ventricular and subventricular zones. J. Neurosci. 20, 5764–5774.
Hebert-Chatelain, E., Desprez, T., Serrat, R., et al., 2016. A cannabinoid link between mitochondria and memory. Nature 539 (7630), 555–559.
Henderson-Redmond, A.N., Nealon, C.M., Davis, B.J., et al., 2020. C-Jun N terminal kinase signaling pathways mediate cannabinoid tolerance in an agonist-specific manner. Neuropharmacology 164, 107847.
Hennou, S., Khalilov, I., Diabira, D., et al., 2002. Early sequential formation of functional GABAA and glutamatergic synapses on CA1 interneurons of the rat foetal hippocampus. Eur. J. Neurosci. 16, 197–208.

Herkenham, M., Lynn, A.B., Little, M.D., et al., 1990. Cannabinoid receptor localization in brain. Proc. Natl. Acad. Sci. U. S. A. 87, 1932–1936.
Herkenham, M., Lynn, A.B., Johnson, M.R., et al., 1991. Characterization and localization of cannabinoid receptors in rat brain, a quantitative in vitro autoradiographic study. J. Neurosci. 11 (2), 563–583.
Hermanns-Clausen, M., Kithinji, J., Spehl, M., et al., 2016. Adverse effects after the use of JWH-210—a case series from the EU spice II plus project. Drug Test. Anal. 8 (10), 1030–1038.
Hermanson, D.J., Gamble-George, J.C., Marnett, L.J., et al., 2014. Substrate-selective COX-2 inhibition as a novel strategy for therapeutic endocannabinoid augmentation. Trends Pharmacol. Sci. 35, 358–367.
Heyser, C.J., Hampson, R.E., Deadwyler, S.A., 1993. Effects of delta-9-tetrahydrocannabinol on delayed match to sample performance in rats. Alterations in short-term memory associated with changes in task specific firing of hippocampal cells. J. Pharmacol. Exp. Ther. 264, 294–307.
Higuera-Matas, A., Ucha, M., Ambrosio, E., 2015. Long-term consequences of perinatal and adolescent cannabinoid exposure on neural and psychological processes. Neurosci. Biobehav. Rev. 55, 119–146.
Hill, M.N., Titterness, A.K., Morrish, A.C., et al., 2010. Endogenous cannabinoid signaling is required for voluntary exercise-induced enhancement of progenitor cell proliferation in the hippocampus. Hippocampus 523, 513–523.
Hill, J.D., Zuluaga-Ramirez, V., Gajghate, S., et al., 2018. Activation of GPR55 increases neural stem cell proliferation and promotes early adult hippocampal neurogenesis. Br. J. Pharmacol. 175 (16), 3407–3421.
Hill, J.D., Zuluaga-Ramirez, V., Gajghate, S., et al., 2019. Activation of GPR55 induces neuroprotection of hippocampal neurogenesis and immune responses of neural stem cells following chronic, systemic inflammation. Brain Behav. Immun. 76, 165–181.
Hillard, C.J., 2015. The endocannabinoid signaling system in the CNS, a primer. Int. Rev. Neurobiol. 125, 1–47.
Hillard, C.J., Manna, S., Greenberg, M.J., et al., 1999. Synthesis and characterization of potent and selective agonists of the neuronal cannabinoid receptor (CB1). J. Pharmacol. Exper. Therap. 289, 1427–1433.
Hoffman, A.F., Oz, M., Yang, R., et al., 2007. Opposing actions of chronic Δ9-tetrahydrocannabinol and cannabinoid antagonists on hippocampal long-term potentiation. Learn. Mem. 14, 63–74.
Horswill, J.G., Bali, U., Shaaban, S., et al., 2007. PSNCBAM-1, a novel allosteric antagonist at cannabinoid CB1 receptors with hypophagic effects in rats. Br. J. Pharmacol. 152 (5), 805–814.
Horti, A.G., Gao, Y., Ravert, H.T., et al., 2010. Synthesis and biodistribution of [11C] A-836339, a new potential radioligand for PET imaging of cannabinoid type 2 receptors (CB2). Bioorg. Med. Chem. 18 (14), 5202–5207.
Howlett, A.C., Abood, M.E., 2017. CB1 and CB2 receptor pharmacology. Adv. Pharmacol. 80, 169–206.
Howlett, A.C., Barth, F., Bonner, T.I., et al., 2002. International Union of Pharmacology. XXVII. Classification of cannabinoid receptors. Pharmacol. Rev. 54, 161–202.
Hsieh, C., Brown, S., Derleth, C., et al., 1999. Internalization and recycling of the CB1 cannabinoid receptor. J. Neurochem. 73, 493–501.
Hsu, K.L., Tsuboi, K., Adibekian, A., et al., 2012. DAGLbeta inhibition perturbs a lipid network involved in macrophage inflammatory responses. Nat. Chem. Biol. 8, 999–1007.
Hu, S.S., Bradshaw, H.B., Chen, J.S., et al., 2008. Prostaglandin E2 glycerol ester, an endogenous COX-2 metabolite of 2-arachidonoylglycerol, induces hyperalgesia and modulates NFkappaB activity. Br. J. Pharmacol. 153, 1538–1549.

Ignatowska-Jankowska, B.M., Ghosh, S., Crowe, M.S., et al., 2014. In vivo characterization of the highly selective monoacylglycerol lipase inhibitor KML29, antinociceptive activity without cannabimimetic side effects. Br. J. Pharmacol. 171, 1392–1407.
Ignatowska-Jankowska, B.M., Baillie, G.L., Kinsey, S., et al., 2015. A cannabinoid CB1 receptor-positive allosteric modulator reduces neuropathic pain in the mouse with no psychoactive side effects. Neuropsychopharmacology 40, 2948–2959.
Ilyasov, A.A., Milligan, C.E., Pharr, E.P., et al., 2018. The endocannabinoid system and oligodendrocytes in health and disease. Front. Neurosci. 12, 733.
Irving, A., Abdulrazzaq, G., Chan, S.L.F., et al., 2017. Cannabinoid receptor-related orphan G protein-coupled receptors. Adv. Pharmacol. 80, 223–247.
Jain, T., Wager-Miller, J., Mackie, K., et al., 2013. Diacylglycerol lipasealpha (DAGLalpha) and DAGLbeta cooperatively regulate the production of 2-arachidonoyl glycerol in autaptic hippocampal neurons. Mol. Pharmacol. 84, 296–302.
Javed, H., Azimullah, S., Haque, M.E., et al., 2016. Cannabinoid type 2 (CB2) receptors activation protects against oxidative stress and neuroinflammation associated dopaminergic neurodegeneration in rotenone model of Parkinson's disease. Front. Neurosci. 10, 321.
Jiang, S., Fu, Y., Williams, J., et al., 2007. Expression and function of cannabinoid receptors CB1 and CB2 and their cognate cannabinoid ligands in murine embryonic stem cells. PLoS One 2, e641.
Jiang, B., Huang, S., de Pasquale, R., et al., 2010. The maturation of GABAergic transmission in visual cortex requires endocannabinoid-mediated LTD of inhibitory inputs during a critical period. Neuron 66 (2), 248–259.
Jimenez-Blasco, D., Busquets-Garcia, A., Hebert-Chatelain, E., et al., 2020. Glucose metabolism links astroglial mitochondria to cannabinoid effects. Nature 583 (7817), 603–608.
Jin, K., Xie, L., Kim, S.H., et al., 2004. Defective adult neurogenesis in CB1 cannabinoid receptor knockout mice. Mol. Pharmacol. 66 (2), 204–208.
Jordan, C.J., Xi, Z.X., 2019. Progress in brain cannabinoid CB(2) receptor research, from genes to behavior. Neurosci. Biobehav. Rev. 98, 208–220.
Kallendrusch, S., Hobusch, C., Ehrlich, A., et al., 2012. Site-specific and time-dependent activation of the endocannabinoid system after transection of long-range projections. PLoS One 7 (3), e33537.
Kamat, S.S., Camara, K., Parsons, W.H., et al., 2015. Immunomodulatory lysophosphatidylserines are regulated by ABHD16A and ABHD12 interplay. Nat. Chem. Biol. 11, 164–171.
Kano, M., 2014. Control of synaptic function by endocannabinoid-mediated retrograde signaling. Proc Japan Acad. Ser. B, Phys. Biol. Sci. 90, 235–250.
Kano, M., Ohno-Shosaku, T., Hashimotodani, Y., et al., 2009. Endocannabinoid-mediated control of synaptic transmission. Physiol. Rev. 89 (1), 309–380.
Katona, I., Sperlagh, B., Sik, A., et al., 1999. Presynaptically located CB1 cannabinoid receptors regulate GABA release from axon terminals of specific hippocampal interneurons. J. Neurosci. 19, 4544–4558.
Katona, I., Urban, G.M., Wallace, M., et al., 2006. Molecular composition of the endocannabinoid system at glutamatergic synapses. J. Neurosci. 26, 5628–5637.
Kawamura, Y., Fukaya, M., Maejima, T., et al., 2006. The CB1 cannabinoid receptor is the major cannabinoid receptor at excitatory presynaptic sites in the hippocampus and cerebellum. J. Neurosci. 26, 2991–3001.
Keimpema, E., Mackie, K., Harkany, T., 2011. Molecular model of cannabis sensitivity in developing neuronal circuits. Trends Pharmacol. Sci. 32, 551–561. https://doi.org/10.1016/j.tips.2011.05.004.
Kelley, B.G., Thayer, S.A., 2004. Delta 9-tetrahydrocannabinol antagonizes endocannabinoid modulation of synaptic transmission between hippocampal neurons in culture. Neuropharmacology 46, 709–715.

Kellogg, R., Mackie, K., Straiker, A., 2009. Cannabinoid CB1 receptor-dependent long-term depression in autaptic excitatory neurons. J. Neurophysiol. 102 (2), 1160–1171.
Kendall, D.A., Yudowski, G.A., 2016. Cannabinoid receptors in the central nervous system, their signaling and roles in disease. Front. Cell. Neurosci. 10, 294.
Khurana, L., Mackie, K., Piomelli, D., et al., 2017. Modulation of CB1 cannabinoid receptor by allosteric ligands. Pharmacology and therapeutic opportunities. Neuropharmacology 124, 3–12.
Kim, J., Alger, B.E., 2004. Inhibition of cyclooxygenase-2 potentiates retrograde endocannabinoid effects in hippocampus. Nat. Neurosci. 7, 697–698.
Kim, J., Isokawa, M., Ledent, C., et al., 2002. Activation of muscarinic acetylcholine receptors enhances the release of endogenous cannabinoids in the hippocampus. J. Neurosci. 22, 10182–10191.
Kinsey, S.G., O'Neal, S.T., Long, J.Z., 2011. Inhibition of endocannabinoid catabolic enzymes elicits anxiolytic-like effects in the marble burying assay. Pharmacol. Biochem. Behav. 98, 21–27.
Kinsey, S.G., Wise, L.E., Ramesh, D., et al., 2013. Repeated low-dose administration of the monoacylglycerol lipase inhibitor JZL184 retains cannabinoid receptor type 1-mediated antinociceptive and gastroprotective effects. J. Pharmacol. Exp. Ther. 345, 492–501.
Koch, M., Varela, L., Kim, J.G., et al., 2015. Hypothalamic POMC neurons promote cannabinoid-induced feeding. Nature 519, 45–50.
Kochman, L.J., dos Santos, A.A., Fornal, C.A., et al., 2006. Despite strong behavioral disruption, Delta 9-tetrahydrocannabinol does not affect cell proliferation in the adult mouse dentate gyrus. Brain Res. 1113 (1), 86–93.
Kozak, K.R., Rowlinson, S.W., Marnett, L.J., 2000. Oxygenation of the endocannabinoid, 2-arachidonylglycerol, to glyceryl prostaglandins by cyclooxygenase-2. J. Biol. Chem. Th275 (43), 33744–33749.
Kreitzer, A.C., Carter, A.G., Regehr, W.G., 2002. Inhibition of interneuron firing extends the spread of endocannabinoid signaling in the cerebellum. Neuron 34, 787–796.
Laaris, N., Good, C.H., Lupica, C.R., 2010. Delta 9-tetrahydrocannabinol is a full agonist at CB1 receptors on GABA neuron axon terminals in the hippocampus. Neuropharmacology 59 (1–2), 121–127.
Lauckner, J.E., Hille, B., Mackie, K., 2005. The cannabinoid agonist WIN55,212-2 increases intracellular calcium via CB1 receptor coupling to Gq/11 G proteins. Proc. Natl. Acad. Sci. U. S. A. 102 (52), 19144–19149.
Laurie, D.J., Wisden, W., Seeburg, P.H., 1992. The distribution of thirteen GABAA receptor subunit mRNAs in the rat brain. III. Embryonic and postnatal development. J. Neurosci. 12, 4151–4172.
Ledent, C., Valverde, O., Cossu, G., et al., 1999. Unresponsiveness to cannabinoids and reduced addictive effects of opiates in CB1 receptor knockout mice. Science 283, 401–404.
Leleu-Chavain, N., Desreumaux, P., Chavatte, P., et al., 2013. Therapeutical potential of CB(2) receptors in immune-related diseases. Curr. Mol. Pharmacol. 6 (3), 183–203.
Leterrier, C., Laine, J., Darmon, M., et al., 2006. Constitutive activation drives compartment-selective endocytosis and axonal targeting of type 1 cannabinoid receptors. J. Neurosci. 26, 3141–3153.
Li, Y., Kim, J., 2015. Neuronal expression of CB2 cannabinoid receptor mRNAs in the mouse hippocampus. Neuroscience 311, 253–267.
Li, X., Hua, T., Vemuri, K., et al., 2019. Crystal structure of the human cannabinoid receptor CB2. Cell 176, 459–467 (e413.r).
Lichtman, A.H., Dimen, K.R., Martin, B.R., 1995. Systemic or intrahippocampal cannabinoid administration impairs spatial memory in rats. Psychopharmacology (Berl) 119, 282–290.

Ligresti, A., De Petrocellis, L., Di Marzo, V., 2016. From phytocannabinoids to cannabinoid receptors and endocannabinoids. Pleiotropic physiological and pathological roles through complex pharmacology. Physiol. Rev. 96, 1593–1659.
Liu, X., Wang, Q., Haydar, T.F., et al., 2005. Nonsynaptic GABA signaling in postnatal subventricular zone controls proliferation of GFAP-expressing progenitors. Nat. Neurosci. 8, 1179–1187.
Liu, J., Wang, L., Harvey-White, J., et al., 2006. A biosynthetic pathway for anandamide. Proc. Natl. Acad. Sci. U. S.A. 103, 13345–13350.
Liu, Q.R., Canseco-Alba, A., Liang, Y., et al., 2020. Low basal CB2R in dopamine neurons and microglia influences cannabinoid tetrad effects. Int. J. Mol. Sci. 21 (24), 9763.
Long, J.Z., Li, W., Booker, L., et al., 2009. Selective blockade of 2-arachidonoylglycerol hydrolysis produces cannabinoid behavioral effects. Nat. Chem. Biol. 5, 37–44.
LoTurco, J.J., Kriegstein, A.R., 1991. Clusters of coupled neuroblasts in embryonic neocortex. Science 252, 563–566.
LoTurco, J.J., Owens, D.F., Heath, M.J., et al., 1995. GABA and glutamate depolarize cortical progenitor cells and inhibit DNA synthesis. Neuron 15, 1287.
Lu, H.C., Mackie, K., 2016. An introduction to the endogenous cannabinoid system. Biol. Psychiatry 79, 516–525.
Ludanyi, A., Hu, S.S., Yamazaki, M., et al., 2011. Complementary synaptic distribution of enzymes responsible for synthesis and inactivation of the endocannabinoid 2-arachidonoylglycerol in the human hippocampus. Neuroscience 174, 50–63.
Luján, M.Á., Castro-Zavala, A., Alegre-Zurano, L., et al., 2018. Repeated cannabidiol treatment reduces cocaine intake and modulates neural proliferation and CB1R expression in the mouse hippocampus. Neuropharmacology 143, 163–175.
Ma, W., Barker, J.L., 1995. Complementary expressions of transcripts encoding GAD67 and GABAA receptor α4, β1, and γ1 subunits in the proliferative zone of the embryonic rat central nervous system. J. Neurosci. 15, 2547–2560.
Ma, L., Jia, J., Liu, X., et al., 2015. Activation of murine microglial N9 cells is attenuated through cannabinoid receptor CB2 signaling. Biochem. Biophys. Res. Commun. 458, 92–97.
Maccarrone, M., 2009. The endocannabinoid system and its manifold central actions. In: Lajtha, A., Tettamanti, G., Goracci, G. (Eds.), Handbook of Neurochemistry and Molecular Neurobiology, Neural Lipids. Springer US, Boston, MA, USA, pp. 385–405.
Maccarrone, M., Fiorucci, L., Erba, F., et al., 2000. Human mast cells take up and hydrolyze anandamide under the control of 5-lipoxygenase and do not express cannabinoid receptors. FEBS Lett. 468 (2–3), 176–180.
Maccarrone, M., Guzmán, M., Mackie, K., et al., 2014. Programming of neural cells by (endo)cannabinoids, from physiological rules to emerging therapies. Nat. Rev. Neurosci. 15 (12), 786–801.
Mackie, K., 2005. Distribution of cannabinoid receptors in the central and peripheral nervous system. Handbook Exp. Pharmacol. 168, 299–325. https://doi.org/10.1007/3-540-26573-2_10.
Maćkowiak, M., Chocyk, A., Markowicz-Kula, K., et al., 2007. Acute activation of CB1 cannabinoid receptors transiently decreases PSA-NCAM expression in the dentate gyrus of the rat hippocampus. Brain Res. 1148, 43–52.
Maneuf, Y.P., Nash, J.E., Crossman, A.R., et al., 1996. Activation of the cannabinoid receptor by [Delta]9-tetrahydrocannabinol reduces [gamma]-aminobutyric acid uptake in the globus pallidus. Eur. J. Pharmacol. 308, 161–164.
Marchalant, Y., Brothers, H.M., Norman, G.J., et al., 2009. Cannabinoids attenuate the effects of aging upon neuroinflammation and neurogenesis. Neurobiol. Dis. 34, 300–307.

Maresz, K., Carrier, E.J., Ponomarev, E.D., et al., 2005. Modulation of the cannabinoid CB2 receptor in microglial cells in response to inflammatory stimuli. J. Neurochem. 95, 437–445.

Marinelli, S., Pacioni, S., Bisogno, T., et al., 2008. The endocannabinoid 2-arachidonoylglycerol is responsible for the slow self-inhibition in neocortical interneurons. J. Neurosci. 28, 13532–13541.

Marinelli, S., Pacioni, S., Cannich, A., et al., 2009. Self-modulation of neocortical pyramidal neurons by endocannabinoids. Nat. Neurosci. 12, 1488–1490.

Marrs, W.R., Blankman, J.L., Horne, E.A., et al., 2010. The serine hydrolase ABHD6 controls the accumulation and efficacy of 2-AG at cannabinoid receptors. Nat. Neurosci. 13 (8), 951–957.

Marshell, R., Kearney-Ramos, T., Brents, L.K., et al., 2014. In vivo effects of synthetic cannabinoids JWH-018 and JWH-073 and phytocannabinoid Δ9-THC in mice, inhalation versus intraperitoneal injection. Pharmacol. Biochem. Behav. 124, 40–47.

Marsicano, G., Lutz, B., 1999. Expression of the cannabinoid receptor CB1 in distinct neuronal subpopulations in the adult mouse forebrain. Eur. J. Neurosci. 11, 4213–4225.

Martini, L., Thompson, D., Kharazia, V., et al., 2010. Differential regulation of behavioral tolerance to WIN55,212-2 by GASP1. Neuropsychopharmacology 35, 1363–1373.

Martin-Moreno, A.M., Reigada, D., Ramirez, B.G., et al., 2011. Cannabidiol and other cannabinoids reduce microglial activation in vitro and in vivo, relevance to Alzheimer's disease. Mol. Pharmacol. 79, 964–973.

Mato, S., Robbe, D., Puente, N., et al., 2005. Presynaptic homeostatic plasticity rescues long-term depression after chronic Delta 9-tetrahydrocannabinol exposure. J. Neurosci. 25, 11619–11627.

Matsuda, L.A., Lolait, S.J., Brownstein, M.J., et al., 1990. Structure of a cannabinoid receptor and functional expression of the cloned cDNA. Nature 346, 561–564.

Matyas, F., Yanovsky, Y., Mackie, K., et al., 2006. Subcellular localization of type 1 cannabinoid receptors in the rat basal ganglia. Neuroscience 137, 337–361.

Matyas, F., Watanabe, M., Mackie, K., et al., 2007. Molecular architecture of the cannabinoid signaling system in the core of the nucleus accumbens. Ideggyogy. Sz. 60, 187–191.

McIntosh, H.H., Song, C., Howlett, A.C., 1998. CB1 cannabinoid receptor, cellular regulation and distribution in N18TG2 neuroblastoma cells. Brain Res. Mol. Brain Res. 53, 163–173.

McPartland, J.M., Glass, M., Pertwee, R.G., 2007. Meta-analysis of cannabinoid ligand binding affinity and receptor distribution, interspecies differences. Br. J. Pharmacol. 152, 583–593.

Mecha, M., Feliú, A., Carrillo-Salinas, F.J., et al., 2015. Endocannabinoids drive the acquisition of an alternative phenotype in microglia. Brain Behav. Immun. 49, 233–245.

Mensching, L., Djogo, N., Keller, C., Rading, S., Karsak, M., 2019. Stable adult hippocampal neurogenesis in cannabinoid receptor CB2 deficient mice. Int. J. Mol. Sci. 20 (15), 3759.

Mereu, G., Fà, M., Ferraro, L., et al., 2003. Prenatal exposure to a cannabinoid agonist produces memory deficits linked to dysfunction in hippocampal long-term potentiation and glutamate release. Proc. Natl. Acad. Sci. U. S. A. 100, 4915–4920.

Michell-Robinson, M.A., Touil, H., Healy, L.M., et al., 2015. Roles of microglia in brain development, tissue maintenance and repair. Brain 138, 1138–1159.

Misner, D.L., Sullivan, J.M., 1999. Mechanism of cannabinoid effects on long-term potentiation and depression in hippocampal CA1 neurons. J. Neurosci. 19, 6795–6805.

Molina-Holgado, E., Vela, J.M., Arévalo-Martín, A., et al., 2002. Cannabinoids promote oligodendrocyte progenitor survival. Involvement of cannabinoid receptors and phosphatidylinositol-3 kinase/Akt signaling. J. Neurosci. 22, 9742–9753.

Molina-Holgado, F., Rubio-Araiz, A., García-Ovejero, D., et al., 2007. CB2 cannabinoid receptors promote mouse neural stem cell proliferation. Eur. J. Neurosci. 25, 629–634.
Montecucco, F., Burger, F., Mach, F., et al., 2008. CB2 cannabinoid receptor agonist JWH-015 modulates human monocyte migration through defined intracellular signaling pathways. Amer. J. Physiol. Heart Circulatory Physiol 294 (3), H1145–H1155.
Morales, P., Hurst, D.P., Reggio, P.H., 2017. Molecular targets of the phytocannabinoids. A complex picture. In: Kinghorn, A.D., Falk, H., Gibbons, S., Kobayashi, J. (Eds.), Phytocannabinoids. vol. 103. Springer Inter. Publishing, Cham, Switzerland, pp. 103–131.
Morgan, D.J., Davis, B.J., Kearn, C.S., et al., 2014. Mutation of putative GRK phosphorylation sites in the cannabinoid receptor 1 (CB1R) confers resistance to cannabinoid tolerance and hypersensitivity to cannabinoids in mice. J. Neurosci. 34, 5152–5163.
Mukhopadhyay, S., Howlett, A.C., 2005. Chemically distinct ligands promote differential CB1 cannabinoid receptor-Gi protein interactions. Mol. Pharmacol. 67 (6), 2016–2024.
Mulder, J., Aguado, T., Keimpema, E., et al., 2008. Endocannabinoid signaling controls pyramidal cell specification and long-range axon patterning. Proc. Natl. Acad. Sci. U. S. A. 105, 8760–8765.
Munro, S., Thomas, K.L., Abu-Shaar, M., 1993. Molecular characterization of a peripheral receptor for cannabinoids. Nature 365 (6441), 61–65.
Murataeva, N., Straiker, A., Mackie, K., 2014. Parsing the players, 2-arachidonoylglycerol synthesis and degradation in the CNS. Br. J. Pharmacol. 171 (6), 1379–1391.
Nakane, S., Oka, S., Arai, S., et al., 2002. 2-arachidonoyl-sn-glycero-3-phosphate, an arachidonic acid-containing lysophosphatidic acid, occurrence and rapid enzymatic conversion to 2-arachidonoyl-sn-glycerol, a cannabinoid receptor ligand, in rat brain. Arch. Biochem. Biophys. 402 (1), 51–58.
Navarrete, M., Araque, A., 2008. Endocannabinoids mediate neuron-astrocyte communication. Neuron 57, 883–893.
Navarro, H.A., Howard, J.L., Pollard, G.T., et al., 2009. Positive allosteric modulation of the human cannabinoid (CB) receptor by RTI-371, a selective inhibitor of the dopamine transporter. Br. J. Pharmacol. 156 (7), 1178–1184.
Navarro, G., Borroto-Escuela, D., Angelats, E., et al., 2018. Receptor-Heteromer mediated regulation of endocannabinoid signaling in activated microglia. Role of CB1 and CB2 receptors and relevance for Alzheimer's disease and levodopa-induced dyskinesia. Brain Behav. Immun. 67, 139–151.
Naydenov, A.V., Horne, E.A., Cheah, C.S., et al., 2014. ABHD6 blockade exerts antiepileptic activity in PTZ-induced seizures and in spontaneous seizures in R6/2 mice. Neuron 83, 361–371.
Nealon, C.M., Henderson-Redmond, A.N., Hale, D.E., et al., 2019. Tolerance to WIN55,212-2 is delayed in desensitization-resistant S426A/S430A mice. Neuropharmacology 148, 151–159.
Neu, A., Foldy, C., Soltesz, I., 2007. Postsynaptic origin of CB1-dependent tonic inhibition of GABA release at cholecystokinin-positive basket cell to pyramidal cell synapses in the CA1 region of the rat hippocampus. J. Physiol. 578, 233–247.
Nguyen, L., Malgrange, B., Breuskin, I., et al., 2003. Autocrine/paracrine activation of the GABAA receptor inhibits the proliferation of neurogenic polysialylated neural cell adhesion molecule-positive (PSA-NCAM+) precursor cells from postnatal striatum. J. Neurosci. 23, 3278–3294.
Nguyen, X.-T.-A., Almushattat, H., Strubbe, I., et al., 2021. The phenotypic Spectrum of patients with PHARC syndrome due to variants in ABHD12. Genes 12, 1404.
Niyuhire, F., Varvel, S.A., Martin, B.R., et al., 2007. Exposure to marijuana smoke impairs memory retrieval in mice. J Pharmacol. Exper. Therap. 322, 1067–1075.

Nogueras-Ortiz, C., Yudowski, G., 2016. The multiple waves of cannabinoid 1 receptor signaling. Mol. Pharmacol. 90, 620–626. https://doi.org/10.1124/mol.116.104539.

Nomura, D., Morrison, B., Blankman, J., et al., 2011. Endocannabinoid hydrolysis generates brain prostaglandins that promote neuroinflammation. Science 334, 809–813. https://doi.org/10.1126/science.1209200.

Nosi, D., Lana, D., Giovannini, M.G., et al., 2021. Neuroinflammation, integrated nervous tissue response through intercellular interactions at the "whole system" scale. Cell 10 (5), 1195.

Nowicky, A.V., Teyler, T.J., Vardaris, R.M., 1987. The modulation of long-term potentiation by delta-9-tetrahydrocannabinol in the rat hippocampus, in vitro. Brain Res. Bull. 19, 663–672.

Nunez, E., Benito, C., Pazos, M.R., Barbachano, A., Fajardo, O., Gonzolez, S., Tolon, R.M., Romero, J., 2004. Cannabinoid CB2 receptors are expressed by perivascular microglial cells in the human brain, an immunohistochemical study. Synapse 53, 208–213.

Nyilas, R., Dudok, B., Urbán, G.M., et al., 2008. Enzymatic machinery for endocannabinoid biosynthesis associated with calcium stores in glutamatergic axon terminals. J. Neurosci. 28 (5), 1058–1063.

Nyiri, G., Cserep, C., Szabadits, E., et al., 2005. CB1 cannabinoid receptors are enriched in the perisynaptic annulus and on preterminal segments of hippocampal GABAergic axons. Neuroscience 136, 811–822.

Ohno-Shosaku, T., Maejima, T., Kano, M., 2001. Endogenous cannabinoids mediate retrograde signals from depolarized postsynaptic neurons to presynaptic terminals. Neuron 29, 729–738.

Ojha, S., Javed, H., Azimullah, S., et al., 2016. β-Caryophyllene, a phytocannabinoid attenuates oxidative stress, neuroinflammation, glial activation, and salvages dopaminergic neurons in a rat model of Parkinson disease. Mol. Cell. Biochem. 418, 59–70.

Oliveira da Cruz, J.F., Robin, L.M., Drago, F., et al., 2016. Astroglial type-1 cannabinoid receptor (CB1), a new player in the tripartite synapse. Neuroscience 323, 35–42.

Onaivi, E.S., Ishiguro, H., Gong, J.P., et al., 2006. Discovery of the presence and functional expression of cannabinoid CB2 receptors in brain. Ann. N. Y. Acad. Sci. 1074, 514–536.

Ossato, A., Canazza, I., Trapella, C., et al., 2016. Effect of JWH-250, JWH-073 and their interaction on "tetrad", sensorimotor, neurological and neurochemical responses in mice. Prog. NeuroPsychopharmacol. Biol. Psych. 67, 31–50.

O'Sullivan, S.E., 2007. Cannabinoids go nuclear, evidence for activation of peroxisome proliferator-activated receptors. Br. J. Pharmacol. 152, 576–582.

Owens, D.F., Kriegstein, A.R., 2002. Is there more to GABA than synaptic inhibition? Natl. Rev. 3, 715–727.

Owens, D.F., Liu, X., Kriegstein, A.R., 1999. Changing properties of GABAA receptor-mediated signaling during early neocortical development. J. Neurophysiol. 82, 570–583.

Palazuelos, J., Aguado, T., Egia, A., et al., 2006. Non-psychoactive CB2 cannabinoid agonists stimulate neural progenitor proliferation. FASEB J. 20, 2405–2407.

Palazuelos, J., Ortega, Z., Díaz-Alonso, J., et al., 2012. CB2 cannabinoid receptors promote neural progenitor cell proliferation via mTORC1 signaling. J. Biol. Chem. 287 (2), 1198–1209.

Parent, J.M., 2003. 2003 injury-induced neurogenesis in the adult mammalian brain. Neuroscientist 9 (4), 261–272.

Pauli, C.S., Conroy, M., Vanden Heuvel, B.D., et al., 2020. Cannabidiol drugs clinical trial outcomes and adverse effects. Front. Pharmacol. 11, 63.

Pereira, A.C., Huddleston, D.E., Brickman, A.M., et al., 2007. An in vivo correlate of exercise-induced neurogenesis in the adult dentate gyrus. Proc. Natl. Acad. Sci. U. S. A. 104 (13), 5638–5643.
Pertwee, R.G., 2005. Pharmacological actions of cannabinoids. Handb. Exp. Pharmacol., 1–51.
Pertwee, R.G., Howlett, A.C., Abood, M.E., et al., 2010. International Union of Basic and Clinical Pharmacology. LXXIX. Cannabinoid receptors and their ligands, beyond CB1 and CB2. Pharmacol. Rev. 62, 588–631.
Pickel, V.M., Shobin, E.T., Lane, D.A., et al., 2012. Cannabinoid-1 receptors in the mouse ventral pallidum are targeted to axonal profiles expressing functionally opposed opioid peptides and contacting N-acylphosphatidylethanolamine-hydrolyzing phospholipase D terminals. Neuroscience 227, 10–21.
Piomelli, D., 2003. The molecular logic of endocannabinoid signalling. Nat. Rev. Neurosci. 4 (11), 873–884.
Pizza, V., Agresta, A., D'Acunto, C.W., et al., 2011. Neuroinflamm-aging and neurodegenerative diseases, an overview. CNS Neurol. Disord. Drug Targets 10 (5), 621–634.
Plaza-Zabala, A., Sierra-Torre, V., Sierra, A., 2017. Autophagy and microglia, novel partners in neurodegeneration and aging. Int. J. Mol. Sci. 18 (3), 598.
Plotkin, M.D., Snyder, E.Y., Hebert, S.C., et al., 1997. Expression of the Na-K-2Cl cotransporter is developmentally regulated in postnatal rat brains, a possible mechanism underlying GABA's excitatory role in immature brain. J. Neurobiol. 33, 781–795.
Potts, A.J., Cano, C., Thomas, S.H.L., 2020. Synthetic cannabinoid receptor agonists. Clin. Toxicol. 58, 82–98.
Poulter, M.O., Barker, J.L., O'Carroll, A.M., et al., 1992. Differential and transient expression of GABAA receptor α-subunit mRNAs in the developing rat CNS. J. Neurosci. 12, 2888–2900.
Poulter, M.O., Barker, J.L., O'Carroll, A.M., et al., 1993. Co-existent expression of GABAA receptor β2, β3 and γ2 subunit messenger RNAs during embryogenesis and early postnatal development of the rat central nervous system. Neuroscience 53, 1019–1033.
Pratt, M., Stevens, A., Thuku, M., et al., 2019. Benefits and harms of medical cannabis: a scoping review of systematic reviews. Syst. Rev. 8, 320.
Prenderville, J.A., Kelly, Á.M., Downer, E.J., 2015. The role of cannabinoids in adult neurogenesis. Br. J. Pharmacol. 172 (16), 3950–3963.
Price, M.R., Baillie, G.L., Thomas, A., et al., 2005. Allosteric modulation of the cannabinoid CB1 receptor. Mol. Pharmacol. 68 (5), 1484–1495.
Price, D.A., Martinez, A.A., Seillier, A., et al., 2009. WIN55,212-2, a cannabinoid receptor agonist, protects against nigrostriatal cell loss in the 1-methyl-4-phenyl-1, 2,3,6-tetrahydropyridine mouse model of parkinson's disease. Eur. J. Neurosci. 29, 2177–2186.
Puighermanal, E., Marsicano, G., Busquets-Garcia, A., et al., 2009. Cannabinoid modulation of hippocampal long-term memory is mediated by mTOR signaling. Nat. Neurosci. 12, 1152–1158.
Puighermanal, E., Busquets-Garcia, A., Maldonado, R., et al., 2012. Cellular and intracellular mechanisms involved in the cognitive impairment of cannabinoids. Philos. Trans. R. Soc. Lond. B Biol. Sci. 367 (1607), 3254–3263.
Ramirez, B.G., Blazquez, C., del Pulgar, T.G., et al., 2005. Prevention of Alzheimer's disease pathology by cannabinoids, neuroprotection mediated by blockade of microglial activation. J. Neurosci. 25, 1904–1913.
Ramirez, S.H., Haskó, J., Skuba, A., et al., 2012. Activation of cannabinoid receptor 2 attenuates leukocyte-endothelial cell interactions and blood-brain barrier dysfunction under inflammatory conditions. J. Neurosci. 32 (12), 4004–4016.

Ranganathan, M., D'Souza, D.C., 2006. The acute effects of cannabinoids on memory in humans, a review. Psychopharmacology (Berl) 188, 425–444.
Razani, B., Woodman, S.E., Lisanti, M.P., 2002. Caveolae, from cell biology to animal physiology. Pharmacol. Rev. 54, 431–467.
Represa, A., Ben-Ari, Y., 2005. Trophic actions of GABA on neuronal development. Trends Neurosci. 28 (6), 278–283.
Rivera, P., Bindila, L., Pastor, A., et al., 2015. Pharmacological blockade of the fatty acid amide hydrolase (FAAH) alters neural proliferation, apoptosis and gliosis in the rat hippocampus, hypothalamus and striatum in a negative energy context. Front. Cell. Neurosci. 9, 1–17.
Robbe, D., Alonso, G., Duchamp, F., et al., 2001. Localization and mechanisms of action of cannabinoid receptors at the glutamatergic synapses of the mouse nucleus accumbens. J. Neurosci. 21, 109–116.
Roche, M., Finn, D.P., 2010. Brain CB_2 receptors, implications for neuropsychiatric disorders. Pharmaceuticals 3 (8), 2517–2553.
Rodrigues, R.S., Ribeiro, F.F., Ferreira, F., et al., 2017. Interaction between cannabinoid type 1 and type 2 receptors in the modulation of subventricular zone and dentate gyrus neurogenesis. Front. Pharmacol. 8, 1–26.
Rodríguez de Fonseca, F., Del Arco, I., Bermudez-Silva, F.J., et al., 2005. The endocannabinoid system, physiology and pharmacology. Alcohol Alcohol. 40 (1), 2–14.
Roland, A.B., Ricobaraza, A., Carrel, D., et al., 2014. Cannabinoid-induced actomyosin contractility shapes neuronal morphology and growth. Elife 3, e03159.
Roloff, A.M., Thayer, S.A., 2009. Modulation of excitatory synaptic transmission by delta(9)-tetrahydrocannabinol switches from agonist to antagonist depending on firing rate. Mol. Pharmacol. 75, 892–900.
Rozenfeld, R., Devi, L.A., 2008. Regulation of CB1 cannabinoid receptor trafficking by the adaptor protein AP-3. FASEB J. 22, 2311–2322.
Rubino, T., Vigano, D., Premoli, F., et al., 2006. Changes in the expression of G protein-coupled receptor kinases and beta-arrestins in mouse brain during cannabinoid tolerance, a role for RAS-ERK cascade. Mol. Neurobiol. 33, 199–213.
Rueda, D., Navarro, B., Martinez-Serrano, A., et al., 2002. The endocannabinoid anandamide inhibits neuronal progenitor cell differentiation through attenuation of the Rap1/B-Raf/ERK pathway. J. Biol. Chem. 277, 46645–46650.
Russo, E., Guy, G.W., 2006. A tale of two cannabinoids. The therapeutic rationale for combining tetrahydrocannabinol and cannabidiol. Med. Hypotheses 66, 234–246.
Saez, T.M.M., Aronne, M.P., Caltana, L., Brusco, A.H., 2014. Prenatal exposure to the CB1 and CB2 cannabinoid receptor agonist WIN 55,212-2 alters migration of early-born glutamatergic neurons and GABAergic interneurons in the rat cerebral cortex. J. Neurochem. 129, 637–648.
Safo, P.K., Cravatt, B.F., Regehr, W.G., 2006. Retrograde endocannabinoid signaling in the cerebellar cortex. Cerebellum 5, 134–145.
Sagredo, O., González, S., Aroyo, I., et al., 2009. Cannabinoid CB2 receptor agonists protect the striatum against malonate toxicity. Glia 57, 1154–1167.
Sang, N., Zhang, J., Chen, C., 2006. PGE2 glycerol ester, a COX-2 oxidative metabolite of 2-arachidonoyl glycerol, modulates inhibitory synaptic transmission in mouse hippocampal neurons. J. Physiol. 572, 735–745.
Sarnataro, D., Grimaldi, C., Pisanti, S., et al., 2005. Plasma membrane and lysosomal localization of CB1 cannabinoid receptor are dependent on lipid rafts and regulated by anandamide in human breast cancer cells. FEBS Lett. 579, 6343–6349.
Savonenko, A.V., Melnikova, T., Wang, Y., et al., 2015. Cannabinoid CB2 receptors in a mouse model of Aβ amyloidosis: immunohistochemical analysis and suitability as a PET biomarker of neuroinflammation. PLoS One 10 (6), e0129618.

Scarante, F.F., Ribeiro, M.A., Almeida-Santos, A.F., et al., 2021. Glial cells and their contribution to the mechanisms of action of cannabidiol in neuropsychiatric disorders. Front. Pharmacol. 11, 618065.
Schiavon, A.P., Bonato, J.M., Milani, H., et al., 2016. Influence of single and repeated cannabidiol administration on emotional behavior and markers of cell proliferation and neurogenesis in non-stressed mice. Prog. Neuropsychopharmacol. Biol. Psychiatry 64, 27–34.
Schlosburg, J.E., Blankman, J.L., Long, J.Z., et al., 2010. Chronic monoacylglycerol lipase blockade causes functional antagonism of the endocannabinoid system. Nat. Neurosci. 13, 1113–1119.
Schmöle, A.C., Lundt, R., Gennequin, B., Schrage, H., et al., 2015a. Expression analysis of CB2-GFP BAC transgenic mice. PLoS One 10, e0145472.
Schmöle, A.C., Lundt, R., Ternes, S., et al., 2015b. Cannabinoid receptor 2 deficiency results in reduced neuroinflammation in an Alzheimer's disease mouse model. Neurobiol. Aging 36 (2), 710–719.
Schuele, L.L., Schuermann, B., Bilkei-Gorzo, A., et al., 2022. Regulation of adult neurogenesis by the endocannabinoid-producing enzyme diacylglycerol lipase alpha (DAGLa). Sci. Rep. 12 (1), 633.
Schurman, L.D., Lu, D., Kendall, D.A., et al., 2020. Molecular mechanism and cannabinoid pharmacology. Handb. Exp. Pharmacol. 258, 323–353.
Shen, M., Thayer, S.A., 1999. Delta 9-tetrahydrocannabinol acts as a partial agonist to modulate glutamatergic synaptic transmission between rat hippocampal neurons in culture. Mol. Pharmacol. 55, 8–13.
Simon, G.M., Cravatt, B.F., 2010. Characterization of mice lacking candidate N-acyl ethanolamine biosynthetic enzymes provides evidence for multiple pathways that contribute to endocannabinoid production in vivo. Mol. Biosyst. 6, 1411–1418.
Simon, A.C., Loverdo, C., Gaffuri, A.L., et al., 2013. Activation-dependent plasticity of polarized GPCR distribution on the neuronal surface. J. Mol. Cell Biol. 5, 250–265.
Simonin, F., Karcher, P., Boeuf, J.J., et al., 2004. Identification of a novel family of G protein-coupled receptor associated sorting proteins. J. Neurochem. 89, 766–775.
Soethoudt, M., Grether, U., Fingerle, J., et al., 2017. Cannabinoid CB 2 receptor ligand profiling reveals biased signalling and off-target activity. Nat. Commun. 8, 13958.
Solas, M., Francis, P.T., Franco, R., et al., 2013. CB2 receptor and amyloid pathology in frontal cortex of Alzheimer's disease patients. Neurobiol. Aging 34, 805–808.
Solbrig, M.V., Fan, Y., Hermanowicz, N., Morgese, M.G., et al., 2010. A synthetic cannabinoid agonist promotes oligodendrogliogenesis during viral encephalitis in rats. Exp. Neurol. 226, 231–241.
Solymosi, K., Köfalvi, A., 2017. Cannabis, a treasure trove or pandora's box? Mini-Rev. Med. Chem. 17, 1–70.
Stanslowsky, N., Jahn, K., Venneri, A., et al., 2017. Functional effects of cannabinoids during dopaminergic specification of human neural precursors derived from induced pluripotent stem cells. Addict. Biol. 22, 1329–1342.
Stella, N., 2010. Cannabinoid and cannabinoid-like receptors in microglia, astrocytes, and astrocytomas. Glia 58, 1017–1030.
Stella, N., Piomelli, D., 2001. Receptor-dependent formation of endogenous cannabinoids in cortical neurons. Eur. J. Pharmacol. 425 (3), 189–196.
Stella, N., Schweitzer, P., Piomelli, D., 1997. A second endogenous cannabinoid that modulates long-term potentiation. Nature 388 (6644), 773–778.
Stewart, R.R., Hoge, G.J., Zigova, T., et al., 2002. Neural progenitor cells of the neonatal rat anterior subventricular zone express functional GABAA receptors. J. Neurobiol. 2002 (50), 305–322.

Straiker, A., Mackie, K., 2005. Depolarization-induced suppression of excitation in murine autaptic hippocampal neurons. J. Physiol. 569, 501–517.
Straiker, A., Wager-Miller, J., Hu, S.S., et al., 2011. COX-2 and fatty acid amide hydrolase can regulate the time course of depolarization-induced suppression of excitation. Br. J. Pharmacol. 164, 1672–1683.
Suárez, I., Bodega, G., Fernández, B., 2002. Glutamine synthetase in brain. Effect of ammonia. Neurochem. Int. 41, 123–142.
Suarez, J., Bermudez-Silva, F.J., Mackie, K., et al., 2008. Immunohistochemical description of the endogenous cannabinoid system in the rat cerebellum and functionally related nuclei. J. Comp. Neurol. 509, 400–421.
Sugiura, T., Kishimoto, S., Oka, S., et al., 2006. Biochemistry, pharmacology and physiology of 2-arachidonoylglycerol, an endogenous cannabinoid receptor ligand. Prog. Lipid Res. 45, 405–446.
Suryavanshi, S.V., Kovalchuk, I., Kovalchuk, O., 2021. Cannabinoids as key regulators of inflammasome signaling. A current perspective. Front Immunol. 11, 613613.
Tait, R.J., Caldicott, D., Mountain, D., et al., 2016. A systematic review of adverse events arising from the use of synthetic cannabinoids and their associated treatment. Clin. Toxicol. 54 (1), 1–13.
Takesian, A.E., Hensch, T.K., 2013. Balancing plasticity/stability across brain development. Prog. Brain Res. 207, 3–34.
Tanimura, A., Yamazaki, M., Hashimotodani, Y., et al., 2010. The endocannabinoid 2-arachidonoylglycerol produced by diacylglycerol lipase alpha mediates retrograde suppression of synaptic transmission. Neuron 65, 320–327.
Tapia, M., Dominguez, A., Zhang, W., et al., 2017. Cannabinoid receptors modulate neuronal morphology and ankyrin G density at the axon initial segment. Front. Cell. Neurosci. 11, 5.
Tappe-Theodor, A., Agarwal, N., Katona, I., et al., 2007. A molecular basis of analgesic tolerance to cannabinoids. J. Neurosci. 27, 4165–4177.
Tham, M., Yilmaz, O., Alaverdashvili, M., 2019. Allosteric and orthosteric pharmacology of cannabidiol and cannabidiol-dimethylheptyl at the type 1 and type 2 cannabinoid receptors. Br. J. Pharmacol. 176, 1455–1469.
Tolón, R.M., Núñez, E., Pazos, M.R., et al., 2009. The activation of cannabinoid CB2 receptors stimulates in situ and in vitro beta-amyloid removal by human macrophages. Brain Res. 1283, 148–154.
Tomas-Roig, J., Wirths, O., Salinas-Riester, G., et al., 2016. The cannabinoid CB1/CB2 agonist WIN55212.2 promotes oligodendrocyte differentiation in vitro and neuroprotection during the cuprizone-induced central nervous system demyelination. CNS Neurosci. Ther. 22, 387–395.
Trazzi, S., Steger, M., Mitrugno, V.M., et al., 2010. CB1 cannabinoid receptors increase neuronal precursor proliferation through AKT/glycogen synthase kinase-3beta/beta-catenin signaling. J. Biol. Chem. 285, 10098–10109.
Triviño, J.J., von Bernhardi, R., 2021. The effect of aged microglia on synaptic impairment and its relevance in neurodegenerative diseases. Neurochem. Int. 144, 104982.
Tsou, K., Brown, S., Sanudo-Pena, M.C., et al., 1998. Immunohistochemical distribution of cannabinoid CB1 receptors in the rat central nervous system. Neuroscience 83, 393–411.
Tsou, K., Mackie, K., Sanudo-Pena, M.C., et al., 1999. Cannabinoid CB1 receptors are localized primarily on cholecystokinin-containing GABAergic interneurons in the rat hippocampal formation. Neuroscience 93, 969–975.
Tsuboi, K., Sun, Y.X., Okamoto, Y., et al., 2005. Molecular characterization of N-acylethanolamine-hydrolyzing acid amidase, a novel member of the choloylglycine hydrolase family with structural and functional similarity to acid ceramidase. J. Biol. Chem. 280, 11082–11092.

Tsuboi, K., Ikematsu, N., Uyama, T., Deutsch, D.G., Tokumura, A., Ueda, N., 2013. Biosynthetic pathways of bioactive N-acylethanolamines in brain. CNS Neurol. Disord. Drug Targets 12, 7–16.
Tsuboi, K., Uyama, T., Okamoto, Y., Ueda, N., 2018. Endocannabinoids and related N-acylethanolamines, biological activities and metabolism. Inflamm. Regen. 38, 28.
Tyzio, R., Represa, A., Jorquera, I., et al., 1999. The establishment of GABAergic and glutamatergic synapses on CA1 pyramidal neurons is sequential and correlates with the development of the apical dendrite. J. Neurosci. 19, 10372–10382.
Uchigashima, M., Narushima, M., Fukaya, M., et al., 2007. Subcellular arrangement of molecules for 2-arachidonoyl-glycerol-mediated retrograde signaling and its physiological contribution to synaptic modulation in the striatum. J. Neurosci. 27, 3663–3676.
van Beek, A.A., Van den Bossche, J., Mastroberardino, P.G., et al., 2019. Metabolic alterations in aging macrophages, ingredients for inflammaging? Trends Immunol. 40 (2), 113–127.
Van Sickle, M.D., Duncan, M., Kingsley, P.J., et al., 2005. Identification and functional characterization of brainstem cannabinoid CB2 receptors. Science 310 (5746), 329–332.
Varma, N., Carlson, G.C., Ledent, C., et al., 2001. Metabotropic glutamate receptors drive the endocannabinoid system in hippocampus. J. Neurosci. 21, RC188.
Varvel, S.A., Lichtman, A.H., 2002. Evaluation of CB1 receptor knockout mice in the Morris water maze. J. Pharmacol. Exp. Ther. 301, 915–924.
Viscomi, M.T., Oddi, S., Latini, L., et al., 2009. Selective CB2 receptor agonism protects central neurons from remote axotomy-induced apoptosis through the PI3K/Akt pathway. J. Neurosci. 29 (14), 4564–4570.
Wagner, J.A., Varga, K., Ellis, E.F., Rzigalinski, B.A., et al., 1997. Activation of peripheral CB1 cannabinoid receptors in haemorrhagic shock. Nature 390, 518–521.
Walsh, K.B., Andersen, H.K., 2020. Molecular pharmacology of synthetic cannabinoids, delineating CB1 receptor-mediated cell signaling. Int. J. Mol. Sci. 21 (17), 6115.
Walter, L., Franklin, A., Witting, A., et al., 2002. Astrocytes in culture produce anandamide and other acylethanolamides. J. Biol. Chem. 277, 20869–20876.
Walter, L., Franklin, A., Witting, A., et al., 2003. Nonpsychotropic cannabinoid receptors regulate microglial cell migration. J. Neurosci. 23, 1398–1405.
Wang, D.D., Kriegstein, A.R., 2009. Defining the role of GABA in cortical development. J. Physiol. 587 (Pt 9), 1873–1879.
Watson, S., Chambers, D., Hobbs, C., et al., 2008. The endocannabinoid receptor, CB1, is required for normal axonal growth and fasciculation. Mol. Cell. Neurosci. 38, 89–97.
Wickert, M., Hildick, K.L., Baillie, G.L., et al., 2018. The F238L point mutation in the cannabinoid type 1 receptor enhances basal endocytosis via lipid rafts. Front. Mol. Neurosci. 11, 230.
Wiley, J.L., Ryan, W.J., Razdan, R.K., Martin, B.R., 1998. Evaluation of cannabimimetic effects of structural analogs of anandamide in rats. Eur. J. Pharmacol. 355 (2–3), 113–118.
Willoughby, K.A., Moore, S.F., Martin, B.R., et al., 1997. The biodisposition and metabolism of anandamide in mice. J. Pharmacol. Exper. Therap. 282 (1), 243–247.
Wilson, R.I., Nicoll, R.A., 2002. Endocannabinoid signaling in the brain. Science 296, 678–682.
Wilson, W.H., Ellinwood, E.H., Mathew, R.J., Johnson, K., 1994. Effects of marijuana on performance of a computerized cognitive-neuromotor test battery. Psychiatry Res. 51, 115–125.
Wise, L.E., Thorpe, A.J., Lichtman, A.H., 2009. Hippocampal CB1 receptors mediate the memory impairing effects of [Delta]9-tetrahydrocannabinol. Neuropsychopharmacology 34, 2072–2080.
Witting, A., Walter, L., Wacker, J., Moller, T., Stella, N., 2004. P2X7 receptors control 2-arachidonoylglycerol production by microglial cells. Proc. Natl. Acad. Sci. U. S. A. 101, 3214–3219.

Wolf, S.A., Bick-Sander, A., Fabel, K., et al., 2010. Cannabinoid receptor CB1 mediates baseline and activity-induced survival of new neurons in adult hippocampal neurogenesis. Cell Commun. Signal 8, 12.

Woodburn, S.C., Bollinger, J.L., Wohleb, E.S., 2021. The semantics of microglia activation, neuroinflammation, homeostasis, and stress. J. Neuroinflammation 18 (1), 258.

Woodhams, S.G., Chapman, V., Finn, D.P., et al., 2017. The cannabinoid system and pain. Neuropharmacology 124, 105–120.

Woodward, D.F., Liang, Y., Krauss, A.H., 2008. Prostamides (prostaglandin-ethanolamides) and their pharmacology. Br. J. Pharmacol. 153, 410–419.

Worob, A., Wenthur, C., 2019. DARK classics in chemical neuroscience. Synthetic cannabinoids (spice/K2). ACS Chem. Nerosci. https://doi.org/10.1021/acschemneuro.9b00586.

Wu, D.F., Yang, L.Q., Goschke, A., et al., 2008. Role of receptor internalization in the agonist-induced desensitization of cannabinoid type 1 receptors. J. Neurochem. 104, 1132–1143.

Wu, C.-S., Zhu, J., Wager-Miller, J., et al., 2010. Requirement of cannabinoid CB(1) receptors in cortical pyramidal neurons for appropriate development of corticothalamic and thalamocortical projections. Eur. J. Neurosci. 32, 693–706.

Wu, C.-S., Morgan, D., Jew, C.P., et al., 2014. Long-term consequences of perinatal fatty acid amino hydrolase inhibition. Br. J. Pharmacol. 171, 1420–1434.

Xapelli, S., Agasse, F., Sardà-Arroyo, L., et al., 2013. Activation of type 1 cannabinoid receptor (CB1R) promotes neurogenesis in murine subventricular zone cell cultures. PLoS One 8 (5), e63529.

Xu, Z., Lv, X.A., Dai, Q., et al., 2016. Acute upregulation of neuronal mitochondrial type-1 cannabinoid receptor and its role in metabolic defects and neuronal apoptosis after TBI. Mol. Brain 9 (1), 75.

Yang, H., Zhang, J., Andreasson, K., et al., 2008. COX-2 oxidative metabolism of endocannabinoids augments hippocampal synaptic plasticity. Mol. Cell. Neurosci. 37, 682–695.

Yoshida, T., Fukaya, M., Uchigashima, M., et al., 2006. Localization of diacylglycerol lipase-alpha around postsynaptic spine suggests close proximity between production site of an endocannabinoid, 2-arachidonoyl-glycerol, and presynaptic cannabinoid CB1 receptor. J. Neurosci. 26, 4740–4751.

Young, A.P., Denovan-Wright, E.M., 2022. The dynamic role of microglia and the endocannabinoid system in neuroinflammation. Front. Pharmacol. 12, 806417.

Young, J.K., Heinbockel, T., Gondré-Lewis, M.C., 2013. Astrocyte fatty acid binding protein-7 is a marker for neurogenic niches in the rat hippocampus. Hippocampus 23, 1476–1483.

Zagzoog, A., Mohamed, K.A., Kim, H., 2020. In vitro and in vivo pharmacological activity of minor cannabinoids isolated from Cannabis sativa. Sci. Rep. 10 (1), 20405.

Zamberletti, E., Rubino, T., 2022. Dos(e) age, role of dose and age in the long-term effect of cannabinoids on cognition. Molecules 27, 1411.

Zawilska, J.B., Wojcieszak, J., 2014. Spice/K2 drugs—more than innocent substitutes for marijuana. Int. J. Neuropsychopharmacol. 17 (3), 509–525.

Zhang, J., Chen, C., 2008. Endocannabinoid 2-arachidonoylglycerol protects neurons by limiting COX-2 elevation. J. Biol. Chem. 283 (33), 22601–22611.

Zhang, H.-Y., Gao, M., Liu, Q.-R., et al., 2014. Cannabinoid CB2 receptors modulate midbrain dopamine neuronal activity and dopamine-related behavior in mice. Proc. Natl. Acad. Sci. U. S. A. 111, E5007–E5015.

Zhang, H.-Y., Gao, M., Shen, H., et al., 2017. Expression of functional cannabinoid CB receptor in VTA dopamine neurons in rats. Addict. Biol. 22, 752–765.

Zhou, Y., Falenta, K., Lalli, G., et al., 2014. Endocannabinoid signalling in neuronal migration. Int. J. Biochem. Cell Biol. 47, 104–108.

Zimmermann, T., Maroso, M., Beer, A., et al., 2018. Neural stem cell lineage-specific cannabinoid type-1 receptor regulates neurogenesis and plasticity in the adult mouse hippocampus. Cereb. Cortex 28 (12), 4454–4471.

Zurolo, E., Iyer, A.M., Spliet, W.G., et al., 2010. CB1 and CB2 cannabinoid receptor expression during development and in epileptogenic developmental pathologies. Neuroscience 170 (1), 28–41.

CHAPTER TWO

Converging mechanisms in ethanol neurotoxicity

Miriam Beatriz Virgolini[a,b,*] and Ricardo Marcos Pautassi[c,d]
[a]Instituto de Farmacología Experimental de Córdoba-Consejo Nacional de Investigaciones Científicas y Técnicas, Córdoba, Argentina
[b]Departamento de Farmacología Otto Orsingher, Facultad de Ciencias Químicas, Universidad Nacional de Córdoba, Córdoba, Argentina
[c]Instituto de Investigación Médica M. y M. Ferreyra, INIMEC-CONICET, Universidad Nacional de Córdoba, Córdoba, Argentina
[d]Facultad de Psicología, Universidad Nacional de Córdoba, Córdoba, Argentina
*Corresponding author: e-mail address: miriam.virgolini@unc.edu.ar

Contents

1. Introduction

Alcohol use disorder (AUD) has been defined as a medical condition in which the individual keeps drinking alcohol despite repeated attempts to stop or control such behavior, which in turn generates negative social,

Advances in Neurotoxicology, Volume 8
ISSN 2468-7480
https://doi.org/10.1016/bs.ant.2022.06.002

occupational, and health consequences (NIAAA-NIH, 2020). The fourth edition of the Diagnostic and Statistical Manual of Mental Disorders (DSM–IV) provided specific criteria for two distinct disorders (alcohol abuse and alcohol dependence), whereas DSM–V integrated them into a single disorder with mild, moderate, and severe sub-classifications.

Although no specific receptor for alcohol (the substance we will also refer to as ethanol, particularly in the context of pre-clinical studies) has been identified, diverse mechanisms have been proposed to explain its toxicity and associated damage. In addition to the liver, the brain appears to be the principal target of the adverse effects generated by ethanol intoxication and withdrawal. Thus, in this chapter, the effects of ethanol toxicity in the brain will be discussed. Readers interested in this topic would like to complement the present chapter with other, highly relevant and comprehensive, publications, such as Pereira et al. (2015), Tizabi et al. (2021), Erickson et al. (2019), Fernández et al. (2019), Mira et al. (2019), Berríos-Cárcamo et al. (2020), and Tapia-Rojas et al. (2017a, 2017b).

In this chapter, the experimental literature on ethanol-induced neurotoxicity is systematically organized by subject, in logical order beginning from basic concepts regarding the relationship among metabolic bioactivation, imbalance in the redox status, mitochondrial dysfunction, and neuroinflammation. The interconnection between these apparently disparate concepts will be unraveled toward the end of the chapter. Additionally, we will discuss some of the animal models commonly used to induce ethanol-induced neurotoxicity or neuronal degeneration, such as the "Majchrowicz" model of binge ethanol intoxication or the administration of ethanol to neonate rats, which models ethanol exposure at late stages of human gestation. Finally, by revising approved and proposed novel pharmacological approaches to treat AUD we will provide insights of the highly complex interconnection among the affected neurotransmitter systems, the altered synaptic functionality, and the cellular components damaged, after the exposure to ethanol insults. Overall, and depending on the level of analysis on which each study is focused, ethanol neurotoxicity can be considered either a mitochondrial disease, posing the mitochondrial as a "master switch of the disease," a neuroinflammatory pathology with the reactive glia as a protagonist, or a redox imbalance disorder, in a scenario of uncertainty concerning the order of precedence of each factor, a sort of *"the chicken or the egg"* scientific controversy.

2. Ethanol metabolism

Once orally consumed, ethanol is incorporated by passive diffusion through the stomach wall (about 20%), with the remaining 80% absorbed through the duodenum and small intestine walls. The elimination occurs primarily through metabolism (95–98%), with small fractions being excreted unchanged in the breath, sweat, and urine, among others (Manzo-Avalos and Saavedra-Molina, 2010). The relevance of the non-metabolic excretion is not to be dismissed. Early studies have shown that rats perceive ethanol's orosensory cues derived from this non-metabolic excretion and associate them with the aversive pharmacological effects of the drug (Molina et al., 1989).

While the effects of alcohol on various tissues depend on its concentration in the blood, ethanol toxicity is associated with its metabolism (Lachenmeier et al., 2009). In the liver, ethanol is reversibly metabolized to acetaldehyde (a class I carcinogen) by alcohol dehydrogenases (ADHs), a family of cytosolic enzymes with high affinity although low capacity for ethanol, thus becoming rapidly saturated. There are six closely related ADHs with a different pattern of expression in the organism. The heterodimeric ADH1A, ADH1B, and ADH1C belong to class I ADH showing high homology and are responsible for most ethanol oxidation in the liver (Km = 0.013–27 mM). On the other side, ADH4 can contribute to ethanol metabolism at intoxicating levels, being ADH7 the only isoform not expressed in the liver and responsible of metabolism in the gastrointestinal tract and brain, contributing to local acetaldehyde formation in these organs (Edenberg and McClintick, 2018).

In addition, there are alternative metabolic pathways catalyzed by peroxisomal catalases (CAT) and microsomal CYP2E1, the latter characterized by a lower affinity for ethanol than ADHs (Km = 0.5–0.8 g/L) and thus with high relevance in chronic alcohol consumers. Alternatively, small amounts of ethanol cross the blood-brain barrier (BBB) and are metabolized in the brain mostly by CAT and to a lesser extent by CYP2E1 (Heit et al., 2013; Zimatkin et al., 2006). These pathways are shown in the following reaction:

$$C_2H_5OH + NAD^+ \xrightarrow{\text{ADH, CAT, CYP2E1}} C_2H_4O + NADH$$

The last, irreversible step, involves the action of the mitochondrial enzyme aldehyde dehydrogenase. ALDH2 is the main ALDH isoform that converts acetaldehyde to acetate in the brain and liver, involving the translocation of the substrate into the mitochondria. There are 19 human ALDHs, being ALDH1A1, ALDH1B1, and ALDH2 3 the most relevant to acetaldehyde metabolism. They are homo tetramers and have broad substrate specificities, with mitochondrial ALDH2 having the highest affinity for acetaldehyde (Km = 0.2 mM) and a high reaction velocity (V_{max} = 280/min), with the highest levels in liver and adipose tissue while in the brain is located in the microvasculature of the BBB. Hypoactivity of ALDH2 and the resultant acetaldehyde accumulation occurs in approximately 45% of Asians that have the inactive ALDH2*2 isoform. Each mammalian ALDH subunit contains 3 domains: the catalytic, the NAD^+ binding, and the oligomerization site. The ALDH2*2 isoform is inactive due to a defect in the cofactor binding site caused by the replacement of glutamic acid by lysine at position 487 (Wang and He, 2018; Zhao and Wang, 2015). In regard to the other isoforms, the ALDH1B1 has 75% of homology with ALDH2, is also located in the mitochondria, but does not play a critical role in acetaldehyde metabolism due to its lower expression and affinity for the substrate (Km = 55 mM). Similarly, the cytosolic ALDH1A1 has a low affinity for acetaldehyde (Km = 180 mM), playing a minor role when ALDH2 is not active (Edenberg and McClintick, 2018; Vasiliou et al., 2004). Additional details about these enzymes and their involvement in cellular damage will be provided later.

Importantly, the limiting step of the ADH/ALDH reaction is the availability of the cofactor NAD^+, which must be re-oxidized after ethanol metabolism in the Complex I of the electron transfer chain (ETC). A small fraction of the ethanol consumed is eliminated by non-oxidative pathways, which involve the conjugation with endogenous substrates such as fatty acids, phospholipids, sulfates or glucuronic acid (Lieber, 2005).

In the liver the first step is catalyzed in a reversible reaction by the family of enzymes ADHs, the majority (60%) of ethanol that arrives to the brain is oxidated by CAT, as follows:

$$CH_3CH_2OH + H_2O_2 \rightarrow CH_3CHO + 2H_2O \text{ (a)}$$
$$H_2O_2 + H_2O_2 \rightarrow 2H_2O + O_2 \text{ (b)}$$

CAT are heme-containing enzymes found in the peroxisomal fraction of the cell that catalyzes the removal of H_2O_2 (reaction *b*, peroxidatic), but can also oxidize alcohol as shown in reaction *a*, considered catalytic (Aebi, 1984). The brain CAT-derived acetaldehyde has been suggested to play a role in the development of tolerance to alcohol, voluntary ethanol consumption, and the positive reinforcing actions of this drug (Deitrich et al., 2006; Israel et al., 2013; Quertemont and Didone, 2006). In this line, we have demonstrated that CAT activation increases ethanol intake in rats evaluated in a two-bottle choice paradigm (Mattalloni et al., 2013, 2017, 2019). Considering the simplicity of its chemical structure it is unlikely that acetaldehyde interacts with a neuronal receptor. Rather, the presence of a carbonyl group promotes a reaction with nucleophilic groups and the formation of adducts (with lipids, nucleic acids and proteins) that, in turn, can activate receptors. Specifically, acetaldehyde reacts with dopamine (DA) and epinephrine to form tetrahydroisoquinolines (TIQs), or to generate tetrahydrocarbolines with serotonin. These adducts activate opioid receptors by inhibiting GABAergic interneurons in the ventral tegmental area, which in turn promote DA release in the nucleus accumbens (NAcb) (Berríos-Cárcamo et al., 2017; Israel et al., 2015; Xie et al., 2012). Recent reports indicate that among TIQs, the condensation product 1-methyl-6, 7-dihydroxy-1,2,3,4-TIQ (salsolinol) was detected in the posterior ventral tegmental area (pVTA) which was postulated to be responsible for a significant increase of DA release in the NAcb shell via μ-opioid receptor stimulation (Bassareo et al., 2021). Interestingly, some TIQs may be associated to ethanol's neurotoxic effects as salsolinol and derivates, while others as higenamine and methylated TIQs may exert neuroprotection (Peana et al., 2019).

3. Neurotransmission and intracellular pathways

It is known that, at high doses, ethanol enhances GABAergic inhibitory activity through a positive allosteric modulation on GABAA receptors, thus facilitating the presynaptic release of GABA and inducing sedative/hypnotic effects. In addition, ethanol inhibits glutamate, the major excitatory neurotransmitter in the brain through both ionotropic (iGluR) and metabotropic (mGluR) glutamatergic receptors affecting the neurotrophic activity and synaptic plasticity, in particular *N*-methyl-D-aspartate receptor (NMDAR)-mediated activity. Importantly, chronic ethanol exposure continuously inhibits, while alcohol withdrawal increases the number of NMDAR at the surface of the postsynaptic membrane, thus promoting the neuronal vulnerability to glutamate-induced excitotoxicity (Mira et al., 2019). It is noteworthy that ethanol effects on extracellular glutamate and thereby excitotoxicity could be also related to the astrocyte glutamate/cysteine exchanger (xCT). The latter regulates extrasynaptic glutamatergic tone and is upregulated in the NAcb in conditions of chronic ethanol intake yet down-regulated during ethanol's withdrawal (Peana et al., 2014). Thus, it could be claimed that the combined effect of ethanol on GABA stimulation and glutamate inhibition is responsible for most of the ethanol's central effects. Additionally, low ethanol concentrations facilitate DA release in the mesolimbic system, particularly among neuronal bodies in the VTA that project to the NAcb. This reward circuitry is activated during the initial alcohol use and during the transition to dependence, yet is inhibited during ethanol withdrawal, probably prompting relapse into ethanol self-administration as a means to restore dopaminergic functionality. Interestingly, the opioid system can activate DAergic neurons in the VTA by inhibiting the inhibitory tone of the GABAergic interneurons, ultimately releasing DA in the NAcb (Fernandes et al., 2017; Wilson and Matschinsky, 2020).

4. Oxidative stress

Normal cellular biological processes of the cell generate both oxygen and nitrogen bioproducts, many of them highly reactive free radicals (namely ROS and NOS, respectively). The imbalance between free radical species and endogenous antioxidants is known as oxidative stress, an undesired event that provokes functional and structural alterations to lipids, proteins, and nucleic acids (Halliwell, 1992). Major ROS include the hydroxyl radical ($^{\bullet}OH$) and superoxide ($O_2^{\bullet -}$) while hydrogen peroxide (H_2O_2) is an intermediate agent in the generation of $^{\bullet}OH$ via the Fenton or Heber-Weiss reactions, which require Fe^{2+} or Cu^{2+} respectively. The $O_2^{\bullet -}$ that rapidly forms in the vicinity of the internal mitochondrial membrane is mainly released into the matrix by Complex I and Complex III and into the intermembrane space by Complex III. Once formed, it can be converted into H_2O_2 and O_2 by the enzyme superoxide dismutase (SOD). Subsequently, CAT and glutathione peroxidases (GPx) convert H_2O_2 to H_2O and O_2, which altogether with SOD, glutathione reductase (GRx), and peroxiredoxins (Prxs), are the main components of the enzymatic antioxidant system, responsible for counteracting the deleterious effects of ROS. Additionally, thioredoxins (Trxs) are redox-sensitive proteins that use NADPH as the cofactor, in clear competence with other enzymes that also use this compound. In addition, the peptide glutathione (GSH) constitutes the main non-enzymatic antioxidant, particularly in the mitochondria protecting this organelle from excessive ROS generation, being used by GPx and Prx6 to catalyze

H_2O_2 reduction. Other natural antioxidants are ascorbic acid, vitamin E, tocopherol, cysteine, Zn^{2+}, Se^{2+}, among others (Halliwell, 1992).

Neurons are inherently susceptible to oxidative damage due to high respiratory turnover, reliance on oxidative phosphorylation reactions, high Fe concentrations, and low levels of antioxidant enzymatic defenses (Takeuchi, 2010). Interestingly, oxidative damage is usually manifested in the nervous system as lipid peroxidation and the formation of toxic protein oxidation products. The cells produce ROS from mitochondrial oxidative phosphorylation and NADPH oxidases (NOx) enzymes, xanthine oxidase XOX), cyclooxygenase 2 (COX2), cytochromes P450 and lipooxygenases. The physiological roles of the NOx enzymes include host defense, inflammation, post-translational protein processing, cell signaling, and regulation of gene expression (Panday et al., 2015). All of the catalytic subunits of the NOx family contain a domain that is characterized by at least six transmembrane domains, a predicted region for flavin adenine dinucleotide (FAD) and a NOx-binding site at the c-terminal. Some of the catalytic subunits of this enzyme require association with other subunit-functioning proteins such as p22phox, p40phox, p47phox, and p67phox and need the small GTPases Rac1 or Rac2. In resting microglia and in macrophages that can infiltrate the brain under certain pathological conditions, the NOx2 isoform localizes to the plasma membrane in cholesterol-enriched membrane microdomains along with p22phox. After its activation, it is internalized and redistributed to an intracellular compartment consisting of numerous vesicles (Belarbi et al., 2017).

Several in vivo or in vitro studies report that excessive alcohol intake (Cederbaum et al., 2009; Collins and Neafsey, 2012; Comporti et al., 2010) or exposure (Boyadjieva and Sarkar, 2013; Das and Vasudevan, 2007) results in increased cell death by apoptosis as well as increased oxidative stress, along with a decrease in glutathione and cellular activity of antioxidant enzymes. Consistent with this, pre-clinical in vivo studies—which tend to focus on the hippocampus—have shown that ethanol affects most of the cellular antioxidant systems by decreasing hippocampal GSH content, increasing nitrosative stress and lipid peroxidation, and inducing the expression of enzymes such as nitric oxide synthase (NOS), COX2 and NOx. Importantly, COX2 signaling participates in the production of prostaglandins and thromboxanes from arachidonic acid that are released from membrane phospholipids, a process mediated by phospholipase A2 (PLA2). This finding indicates that part of the effects of ethanol on redox imbalance is mediated by this drug altering PLA2 signaling, but also by producing changes in the lipid content and phospholipid composition of the cellular membranes, as will be discussed below. On the other hand, there are evidence that ethanol upregulates gp91phox, the catalytic subunit of NOx, an inducible enzyme responsible for the formation of the $O_2^{\bullet -2}$ (Mira et al., 2019). A thorough discussion of the relevance of these enzymes in neuroinflammatory processes will be presented in the following sections.

In conclusion, the simultaneous presence of the **following events in the brain and liver ethanol metabolism creates a figurative "perfect physiological storm," ultimately leading to an oxidative state imbalance** (Virgolini et al., 2019): (1) defects in cell membranes; (2) acetaldehyde production; (3) changes in the ratio $NADH/NAD^+$; (4) induction of brain and liver CYP2E1; (5) increase in Fe^{2+} concentrations; (6) mitochondrial damage with a decrease in ATP production, and 7) reduction of GSH levels in the mitochondria. All these events contribute to the generation of aldehydes such as 4-hydroxynonenal (4-HNE) and malonyl dialdehyde (MDA), along with the DA metabolite 3,4-dihydroxyphenylacetaldehyde (DOPAL) are substrates of the ALDH family of enzymes.

4.1 Defects in cell membranes

Ethanol's amphipathic characteristics allow passive diffusion across the lipid bilayer cells, reaching almost all tissues in a short time. At the same time, ethanol can change the structural composition of any type of membrane, including the plasma membrane and the organelle's membranes. Moreover, the membrane fluidity is also affected inducing alterations in proteins and lipid environment, modifying several ion channels, enzymes, and receptors, thereby influencing their functions and associated biological processes (Goldstein, 1986).

In particular, the BBB is a physical barrier formed by the capillaries, astrocytes, efflux proteins, endothelial cells, and tight junctions that interacts with perivascular pericytes, microglial cells, astrocytes, and neurons that, together form the neurovascular unit (Abbott et al., 2006). Considering that the BBB is critical for the communication between the peripheral immune system and the brain, it has been demonstrated that binge alcohol drinking (defined in humans as the ingestion of 4 or 5 standard drinks in a single and usually brief occasion, for women and men respectively) compromises BBB function through inflammatory-related mechanisms by cytokine release and the toll-like receptor 4 (TLR4) signaling pathway (Vore and Deak, 2022). Excessive alcohol consumption is responsible for brain injury, white matter loss, reduced brain volume, and neuronal loss associated with the BBB (Pimentel et al., 2020).

4.2 Acetaldehyde production

Acetaldehyde is a highly reactive compound that can interact with thiol and amino groups of amino acids in proteins. The formation of acetaldehyde adducts with proteins may cause inhibition of their function or immune responses. The ALDHs enzymes are important for removing acetaldehyde and other toxic aldehydes, including biogenic and lipid peroxidation–derived aldehydes. Effective removal of acetaldehyde is important to prevent cellular toxicity and also to keep the oxidation of ethanol at adequate levels, as acetaldehyde is a product that likewise inhibits ALDH. The ALDH1 family can oxidize retinal to retinoic acid, thus the possibility that high levels of acetaldehyde compete with retinal for oxidation by ALDH1 may be taken into consideration. Alternatively, damage to the mitochondria may also be mediated by acetaldehyde, a mechanism associated with the generation of fatty liver (Comporti et al., 2010).

There are reports of hepatic ALDH2 inactivation by ROS or RNS through post-translational modifications (i.e., S-nitrosylation, tyrosine nitration) in an oxidant environment (Song et al., 2011). In addition, another cause of ALDH inactivation comes from adducts formation with toxic aldehydes including 4-HNE, MDA, and acetaldehyde, revealing the essential role of oxidative/nitrosative stress in regulating ALDH2 functionality. Importantly, the inhibition mechanism of 4-HNE to ALDH2 is not only mediated by the formation of 4-HNE adducts with the active site Cys302 of ALDH2, but also by the promotion of c-Jun N-terminal kinase (JNK) mediated ALDH2 phosphorylation (Doorn et al., 2006). Finally, hypoactivity of ALDH2 by the presence of the inactive ALDH2*2 isoform may be yet another source of mitochondrial damage (Hao et al., 2018).

On the other hand, a recent report (Cui et al., 2019) demonstrated that acetaldehyde is cytotoxic to HT22 cells and primary cultured cortical neuronal cells, inducing apoptosis by upregulating the expression of caspase family proapoptotic proteins that increase intracellular Ca^{2+} and ROS levels. In addition, the acetaldehyde-mediated increase in NOx, XOX, or NOs activity and expression is regulated by transcriptional process that produces ROS, NO, and peroxynitrite leading to lipid peroxidation, neuroinflammation and CNS neurodegeneration (Haorah et al., 2008). Moreover, acetaldehyde accumulation generates lipid peroxidation, giving rise to end products that damage the membrane such as MDA and the 4-hydroxyalkenals (4-HAL). Thus, acetaldehyde is an important generator of ROS and provokes mitochondrial dysfunction by compromising its own metabolism, which ultimately leads to an increase in acetaldehyde, generating a self-perpetuating vicious circle. Depending on their composition, these adducts can be classified into four large groups (Díaz-Soto and Calderín-Miranda, 2020):

1. Adducts derived by the formation of bonds between acetaldehyde and lysine residues or amino groups of different proteins. These links can be covalent or non-covalent, which determines the stability of the formed adduct.
2. Some aldehyde products derived from lipid peroxidation, such as MDA or 4-HNE, can also form adducts with proteins.
3. The formation of mixed or hybrid adducts, which are those formed by the binding of acetaldehyde and MDA to the same substrate.
4. Radical hydroxyethyls formed in the oxidation of ethanol can form adducts with proteins in the presence of Fe atoms. The attachment of aldehyde groups to biomolecule results in a structural alteration and consequently, a functional modification of these substrates. Thus, after the binding with acetaldehyde molecules, the activity of some enzymes is reduced or even inhibited due to the formation of adducts.

4.2.1 Changes in the NADH/NAD$^+$ ratio

The level of mitochondrial ROS production depends on the easiness of the electron flow in the ETC. When it is highly reduced, the possibility that an electron moves from O_2 to form $O_2^{\bullet-}$ increases (Meyer et al., 2018). Furthermore, the inhibition of the respiratory chain by xenobiotics or by increased NADH due to low demand for ATP and consequent low respiration rate will increase the NADH/NAD$^+$ ratio and will lead to the formation of $O_2^{\bullet-}$. In contrast, in most situations in which the mitochondria breathe normally and the NADH/NAD$^+$ ratio is relatively low, only small amounts of $O_2^{\bullet-}$ are produced (Murphy, 2009). Additionally, the imbalance in the relationship between NADH and NAD$^+$ is responsible for the generation of ROS in several processes such as glycolysis, lipooxidative damage, and sirtuin-mediated protein deacetylation. Thus, both the Complex I and alpha-ketoglutarate dehydrogenases will produce more ROS when the NADH/NAD$^+$ ratio is high; and in turn, the increase in ROS would compromise the ability of Complex I to oxidize NADH, causing a vicious circle (Stefanatos and Sanz, 2011).

Because the ADHs and ALDH2 reactions reduce NAD$^+$ to NADH, the cellular NADH/NAD$^+$ redox ratio is reduced as a consequence of ethanol metabolism. This reduction has profound effects on other metabolic pathways that require NAD$^+$ or are inhibited by NADH such as glycolysis, citric acid cycle, pyruvate dehydrogenase, fatty acid oxidation or gluconeogenesis (Dashty, 2013).

Thus, in the first step of ethanol oxidation, the ADH reactions occur in the cytosol, reducing the cytosolic NADH/NAD$^+$ redox ratio, as shown in the following reaction:

$$\text{Pyruvate} + \text{NADH} \xrightarrow{\text{LDH}} \text{Lactate} + \text{NAD}^+$$

On the other side, the ALDH2 reaction in the mitochondria, the NADH/NAD$^+$ redox ratio is reduced and reflected by the β-hydroxybutyrate/acetoacetate ratio:

$$\text{Acetoacetate} + \text{NADH} \xrightarrow{\beta\text{-hydroxybutyrate dehydrogenase}} \beta\text{-Hydroxybutyrate} + \text{NAD}^+$$

On this basis, it can be concluded that even at relatively low circulating ethanol levels, there is a substantial reduction in cytosolic NAD$^+$ in the liver, placing the mitochondria in a challenging situation with deleterious consequences to the cell (Dashty, 2013).

4.2.2 Induction of brain and liver CYP2E1

The CYP-P450s family of heme enzymes was initially involved in the oxidation of many endobiotic (steroids, fat-soluble vitamins) in a process that was later hijacked by xenobiotics such as drugs, chemical carcinogens, and industrial chemicals among others. The highest levels of CYP-P450 are in the liver, mainly in the endoplasmic reticulum (microsomal fraction) but some are also in mitochondria (Hartman et al., 2017). The CYP2E1 isoform has the highest activity to oxidize ethanol to acetaldehyde (Km = 10 mM), tenfold higher than the Km of ADH for ethanol. In this scenario, at low ethanol concentrations, CYP2E1 may account for about 10% of the total alcohol-oxidizing capacity of the liver and about 20% for the brain (Heit et al., 2013). However, in view of its higher Km, the relevance of CYP2E1 in ethanol oxidation increases as blood alcohol concentrations increase, playing an important role in the metabolism after chronic consumption. In this situation the expression of the CYP2E1 enzyme increases in the brain, which may contribute to the augmented production of ROS in chronic alcohol consumers. This, in turn, promotes neurotoxicity by protein oxidation and enzyme inactivation; as well as production of reactive lipid aldehydes, such as MDA and 4-HNE that cause mitochondrial dysfunction and apoptosis (Wu and Cederbaum, 2005).

4.2.3 *Increase in Fe^{2+} concentrations*

Both, Fe^{2+} and Fe^{3+} are the main forms of the metal in the cells, the first is also called free/labile Fe which servs as a catalyzer and cofactor due to its oxidation capability and high solubility, whereas most Fe is transported and stored in the Fe^{3+} stable form. The redox interconversion between each status contributes to biochemical reactions. The mitochondria are posed in the center of Fe homeostasis because they confer compartmentalization which is key to strictly regulating its cellular levels. It has been pointed out that an excess of Fe induces fragmented mitochondrion due to excessive fission, an event that is associated to neuronal death triggered by oxidative stress, while its deficiency induces a fussed/aggregated mitochondrion (Upadhyay and Agarwal, 2020).

Redox unique Fe properties allow an efficient electron transference in the mitochondria, which is translated into an energetic supply for several biological reactions. However, these Fe reactive properties can also promote the generation of ROS. Moreover, due to redox reactions in the presence of oxygen, Fe is continuously cycling between Fe^{2+} and Fe^{+3}. Due to still unknown reasons, with age and under conditions of Fe overload, mitochondrial redox-active Fe concentration (free Fe) raises, generating $^{\bullet}OH$ damaging proteins, nucleic acids, and membrane phospholipids. The excess of free Fe is involved in the Fenton reaction where Fe^{2+} is oxidated to Fe^{3+} to catalyze H_2O_2 conversion to the radical $^{\bullet}OH$ and hydroxyl anion (^{-}OH). The excess of free ferric Fe leads to the restoration of reduced ferrous Fe and to GSSG depleting GSH.

Another source of free radicals is located in the dopaminergic cells and stems from non-enzymatic DA oxidation, which is mediated by the Fe redox reaction. This results in the production of semiquinones and H_2O_2 at the level of mitochondrial Complex I of the ETC (Carocci et al., 2018). It is important to stress that the mitochondrial insult induced by different environmental and endogenous toxic agents can both initiate and spread Fe release from non-toxic sites (as the Fe-bounded proteins) which can trigger or accelerate neurodegeneration. The Fe—S clusters present in mitochondrial Complex I and III can be attacked by $O_2^{\bullet-}$ releasing free ferrous cations that participate in the Fenton reaction nourishing a general prooxidant cycle. The pair Fe^{2+}/Fe^{3+} can as well stimulate peroxidation of membrane lipids, a scenario that intensifies oxidative stress contributing to the mitochondrial death through the formation of the mitochondrial permeability pore (mPTP) that leads to the oxidated fatty acid rupture and to series of bioproducts including 4-HNE and MDA (Farina et al., 2013).

The preceding statement supports the hypothesis that the mitochondrial lipid oxidation is a critical step in ferroptosis, a cellular death pathway initiated by a disruption of the antioxidant defenses dependent on GSH and GPx4, which results in lipid peroxidation and eventual cell death (Dixon et al., 2012). Indeed, alcohol metabolism produces Fe overload and lipid peroxidation, suggesting that there is another form of cell death in alcohol-related diseases. While mild alcohol consumption is shown to increase Fe stores, high mortality from alcoholic cirrhosis is associated with oxidative damage caused by Fe^{2+} overload. Overall, it is clear that ferroptosis is engaged in the mechanism of alcohol-induced liver injury, and probably is also present in brain tissue (Liu et al., 2020).

4.2.4 Mitochondrial damage

The degree of oxidative stress in the cells depends both on the formation of ROS (and NOS) and on the cellular defenses that prevent their accumulation, a balance that largely depends on the mitochondria, which in turn is severely affected by oxidative stress. This may represent a vicious cycle, leading to more damage in these organelles, with the final fate of the cell depending on their capacity for repair.

The mitochondria are dynamic organelles that regulate the production of ATP, redox balance, and calcium homeostasis in the neuron, playing a key role in the modulation of cell death, intracellular signaling pathways, and many other functions. They conform the powerhouse of the cell as a consequence of the coupling of the flow of electrons through the respiratory complexes I–IV with oxidative phosphorylation. The ETC is also the main source of ROS due to an incomplete reduction of O_2 from Complex I and Complex III. Besides the nuclear DNA (nDNA), the cells contain mitochondrial DNA (mtDNA) that consists of 13 structural genes that encode the subunits of the ETC and ATP synthase and is highly vulnerable to free radical attack.

In terms of mitochondrial control mechanisms, there are two main events that occur under physiological or pathological conditions: (a) mitochondrial biogenesis, fusion, and fission and (b) mitophagy (Murata et al., 2020). The disruption of fission/fusion processes is associated with pathological processes and neurodegeneration. Mitophagy is a specific form of autophagy that removes damaged mitochondria and reduces the organelle mass in the face of stressful events, promoting cell survival (Meyer et al., 2017). In addition, being acetylation an important posttranslational modification that regulates protein functionality, the deacetylases known as sirtuins have been implicated in the regulation of mitochondrial number, turnover, and activity (Khanna et al., 2019). In relation to ethanol, these deacetylases compete with ADH and ALDH2 for the cofactor NAD+. Thereby, when NAD^+ levels are reduced, SIRT1 activity decreases, leading to mitochondrial dysfunction with adverse consequences for the cellular balance of the cell (French, 2016).

Importantly, mitochondrial oxidative damage can affect proteins and DNA, and promotes membrane permeabilization. This results in the release of proteins to the intermembrane space such as cytochrome *c* (cit c), the translocation of the pro-apoptotic factor bax, which forms a complex with the voltage-dependent anion channel (VDAC) producing the activation of caspases and the mitochondrial apoptotic pathway. Additionally, the ATP synthase, an enzyme of the mitochondrial respiratory chain involved in the generation of ATP, is highly sensitive to ROS-induced oxidative damage due to its proximity to the inner mitochondrial membrane where most of the ROS are produced (Nassir and Ibdah, 2014). Furthermore, the polyunsaturated fatty acids of the mitochondrial membrane are very susceptible to lipid peroxidation. This event has a primary role in the redox state of the cell via the production of toxic aldehydes such as MDA and 4-HNE, which can form adducts with nucleic acids, proteins, amino acids, and polypeptides, leading to enzyme inactivation, mitochondrial dysfunction, and ultimately cell death (Murata et al., 2020). On the other hand, ROS can also promote the opening of the permeability transition pore, leading to the permeabilization of the internal mitochondrial membrane (Manzo-Avalos and Saavedra-Molina, 2010; Moro, 2020). Overall, ethanol increases mitochondrial ROS production, alters mitochondrial dynamics, affects mitochondrial bioenergetics, impairs calcium buffering, and may finally induce cell death by the opening of the mitochondrial permeability transition pore (Tapia-Rojas et al., 2017a,b, 2018, 2019).

4.2.5 Reduction of GSH levels in the mitochondria

GSH is a tripeptide that plays a key role in the mitochondrial defense against ROS. It is formed by the amino acids glycine, cysteine, and glutamic acid. One of these amino acids, cystine, is internalized by the xCT antiporter and reduced into cysteine, the rate-limiting precursor of GSH.

Recent studies showed that administration of the GSH precursor *N*-acetylcysteine (NAC) to rats that chronically consumed ethanol reversed alcohol-induced brain oxidative stress. This occurred by NAC restoring the GSSG/GSH ratio and abolishing neuroinflammation. Moreover, at the behavioral level, the administration of NAC to rats markedly inhibited chronic ethanol intake, partially prevented relapse binge drinking (Quintanilla et al., 2016, 2018), and completely reduced ethanol self-administration (Lebourgeois et al., 2018, 2019). Moreover, it has been suggested that NAC acts via Nrf2 to reduce the glutamatergic tone by inhibition of the presynaptic metabotropic mGlu2/3 receptor. The mechanism underlying this effect involves the activation of the xCT cystine-glutamate antiporter and the increase of cystine as a substrate for the antiporter (Shih et al., 2003; Zhou et al., 2018).

Notably, the Nrf2 system activates the transcription of several antioxidant enzymes, including γ-glutamyl-cysteine synthetase, the rate-limiting enzyme in the synthesis of GSH, along with the availability of intracellular cysteine constitutes the rate-limiting factor in the synthesis of GSH. In this line, Quintanilla et al., 2020 recently observed a reduction in ethanol intake after NAC treatment, which was accompanied by the activation of the hippocampal Nrf2-ARE antioxidant signaling pathway, which was prevented by inhibition of the xCT cystine-glutamate transporter and by inhibition of the glutamate presynaptic metabotropic mGlu2/3 receptor. Consistent with these results, chronic ethanol intake in female rats induced changes in hippocampal astrocyte morphology of the glial fibrillary acidic protein (GFAP) and microglial (IBA1, a calcium-binding protein found in microglia and macrophages) density that were accompanied by increases in oxidative stress, determined as the ratio of oxidized/reduced glutathione (GSSG/GSH), all effects that were ameliorated by NAC (Israel et al., 2021).

4.3 Neuroinflammation and oxidative stress

It is now recognized that persistent activation of the neuroimmune system by alcohol use contributes to alcohol-induced neurotoxicity, and ultimately to the development of AUD (Crews et al., 2011, 2015, 2017; Crews and Vetreno, 2016; Meredith et al., 2021). Henriques et al., 2018). As detailed in previous sections, increased free radical production and low antioxidant levels accompanied by mitochondrial dysfunction are hallmarks of alcohol-induced neurotoxicity (Crews et al., 2000). Thus, under chronic or binge alcohol exposure, redox balance is lost, resulting in mitochondrial damage and increased oxidative and nitrosative stress, which in the case of the microglia triggers a broad spectrum of inflammatory mediators (Mayfield et al., 2013; Song et al., 2014).

Neuroimmune-related changes are associated to all brain cell types: microglia, astroglia, and neurons. In response to an insult, the microglia releases cytokines, an event that precedes astrocyte reactivity and drug-induced neurotoxicity. Glial cells, in particular astrocytes are, key players in antioxidant defenses and metabolic homeostatic balance. During the early stages of alcohol consumption, astrocytes prevent ethanol-induced neuronal death by maintaining GSH homeostasis, whereas in chronic alcohol use, the drug may impair overall glial function. At these stages, there are reports of an altered antioxidant and metabolic connection between neurons and astrocytes, along with the impairment of the neuroimmune function of microglia (Adermark and Bowers, 2016).

Microglia, the resident macrophage-like innate immune cells of the CNS, undergoes functional activation that ranges from a proinflammatory and destructive state: M1, characterized by the release of cytokines such as TNF-α, IL-1b, and IL-6 and producing ROS through increased expression of inducible NOS (iNOS) and NOx, to an anti-inflammatory and reparative M2 state (Beynon and Walker, 2012). Alcohol has been linked to a hyperbranched state characterized by modest morphological changes (M1 type) and an increased release of cytokines and other signaling molecules, increasing neurodegeneration and loss of neurogenesis, which has important consequences for the pathogenesis of alcohol-associated disorders (Qin et al., 2008a,b; Crews et al., 2017).

Moreover, both prolonged ethanol withdrawal and reinstatement are associated to oxidative damage and neuroinflammatory processes, events that remained altered after ethanol consumption as previously reported (Fernández-Rodríguez et al., 2022). Minocycline, a tetracycline antibiotic (Plane et al., 2010) prevents ethanol-induced-microglial activation and reduces alcohol self-administration (Agrawal et al., 2011; Qin and Crews, 2012). Chronic alcohol exposure features pro-inflammatory gene induction and microglial activation through TLR stimulation, cytokine production, and NF-kB (nuclear factor kappa-light-chain-enhancer of activated B cells) signal, an important transcription factor involved in inflammatory responses that promotes the expression of proinflammatory cytokines, including tumor necrosis factor α (TNFα), interleukin-1β (IL-1β), IL-6, and monocyte chemoattractant protein-1 (MCP-1) (Guerri and Pascual, 2019). A recent report indicates that ethanol causes the release of inflammation-promoting microvesicles, a class of extracellular vesicles whose blockade prevents the pro-inflammatory signaling by ethanol, making them a target for the development of new pharmacological treatments for AUD (Crews et al., 2021).

These microvesicles are secreted by cells into the extracellular space and can transmit intercellular signals by delivering proteins, metabolites, and nucleic acids to recipient cells. Moreover, exosomes derived from mesenchymal stem cells were found to reduce oxidative stress, neuroinflammation, ethanol consumption, and relapse when administered intranasally (Ezquer et al., 2019a,b).

On the other hand, it was recently reported that female and male mice differ in ethanol's effects on inflammatory-related plasma miRNAs extracellular vesicle profiles, with females being more vulnerable than males to the inflammatory effects of ethanol exposure (Ibáñez et al., 2020). Interestingly, it seems that fenofibrate and tesaglitazar -both peroxisome proliferator-activated receptor (PPAR) agonists- inhibit the expression of inflammatory cytokines and the release of several pro-inflammatory and pro-angiogenic enzymes (e.g., iNOS), and mediators (e.g., NO oxide and TNF-α) (Strosznajder et al., 2021). These ligands have been shown to decrease ethanol consumption and ethanol-related behaviors (Blednov et al., 2015; Haile and Kosten, 2017), to be associated to high blood acetaldehyde levels (Karahanian et al., 2014), and to also regulate the expression of brain genes related to ethanol drinking behavior (Ferguson et al., 2014).

The scenario of cytokines and chemokines release, microglial activation and ROS formation contributes to increased production of inflammatory mediators via NF-κB signaling converting oxidative stress in a remarkable upstream actor in the signaling cascade of inflammation. In this regard, Qin and Crews (2012) demonstrated that mice chronically exposed to alcohol showed increased NOx expression, microglial activation, and cell death in cortical and hippocampal brain regions. In addition, NOx inhibition decreased oxygen, microglial activation, and cell death in the same brain regions during alcohol administration, directly linking ROS production and microglial activation to alcohol-induced neurotoxicity. It can be concluded that the oxidative environment along with the neuroinflammation state generates a self-perpetuating cycle in individuals with alcohol use disorders that is produced by some of all of these five key mechanisms (Berríos-Cárcamo et al., 2020):

4.3.1 Dopamine oxidation

The neurotransmitter DA is synthesized in the presynaptic button to be stored in vesicles until released in the synaptic extracellular cleft. In the synaptic space, the transmitter can interact with postsynaptic DA receptors or regulatory presynaptic DA autoreceptors. The signal is interrupted by DA transporter (DAT)-mediated uptake or oxidative deamination by MAO to form DOPAL which is in most part oxidized to 3,4-dihydroxyphenylacetic acid (DOPAC) by ALDH. In addition, the catechol-O methyltransferase

(COMT) converts the DA present in the synaptic space to homovanilic acid (HVA). Excessive concentrations of cytoplasmic DA can be neurotoxic as a result of ROS formation product of its metabolism either by autooxidation or through the action of the MAO enzyme, leading to oxidative stress and apoptosis (Banerjee et al., 2014). Importantly, MAO converts DA to the intermediate metabolite DOPAL generating H_2O_2 via Haber–Weiss/Fenton reaction to generate the highly toxic $OH^{\bullet}$, while ALDH converts DOPAL to DOPAC (Goldstein, 2020). On the other hand, DA auto-oxidation may occur in the extracellular medium, leading to the generation of $O^{\bullet-}$ and toxic quinones, including aminochrome (Banerjee et al., 2014; Segura-Aguilar, 2019). Furthermore, $O^{\bullet-}$ reacts with NO by neuronal NOS (nNOS) to form the highly toxic peroxynitrite ($ONOO^-$) and other oxidation products that may damage lipids, proteins, nDNA, and mtDNA, affecting cell survival (Surendran and Rajasankar, 2010).

4.3.2 Cystine/glutamate transport system inhibition

The astrocytic xCT antiporter acts interchanging one cystine molecule to the intracellular space and one glutamate molecule to the extracellular space (Baker et al., 2002; Bridges et al., 2012), promoting glutathione synthesis, increasing the antioxidant protection against pro-oxidant drugs of abuse, and maintaining glutamate homeostasis. As mentioned, chronic ethanol consumption and withdrawal from chronic ethanol consumption increase and decrease central xCT expression, respectively (Peana et al., 2014). Moreover, it was recently demonstrated that individual susceptibility to the extinction and reinstatement of ethanol-induced conditioned place preference—a behavioral measure of drug-induced reinforcement—is related to alterations in xCT levels. Specifically, mice that have persistent conditioning to ethanol have a lower expression of xCT in limbic areas, which could lead to a decreased extrasynaptic glutamate level, and in consequence a reduced inhibitory tone of the mGluR2/3 receptors. This, in turn, increases the release of glutamate into the synapse and produced greater activation of ionotropic receptors in the post-synaptic neuron, a scenario that promotes excitotoxicity (Amaral et al., 2021).

4.3.3 Mitochondrial dysfunction

The alteration of mitochondrial functionality has been associated to neuroinflammatory processes triggering a vicious cycle that ends in neuronal death. As mentioned, microglial activation induces a neuroinflammatory response that involves the release of inflammatory mediators such as cytokines and chemokines, which can trigger oxidative and nitrosative stress. This scenario may activate additional microglial cells and cause damage to key components of the mitochondria, such as mtDNA and enzymes that conform to the mitochondrial respiratory chain. Thus, the interaction between neuroinflammatory processes and mitochondria may result in the generation of a self-renewing vicious cycle that ultimately leads to neuronal death (Di Filippo et al., 2010).

In this context, the mitochondria have been implicated in the activation of the inflammasome complex, a cytosolic multiprotein assembly that through the activity of caspase-1 activates proinflammatory cytokines and promotes an inflammatory form of cell death named pyroptosis, which promotes disruption of the BBB and ROS production. It has been reported that mitochondrial ROS, oxidized mtDNA and cardiolipin leakage mediate activation of inflammasomes, although the exact role of mitochondria in this process is still largely unknown. Moreover, the inflammasome activation depends on the interaction between the endoplasmic reticulum and the mitochondria, with mitochondrial Ca^{2+} imbalance also resulting in inflammasome activation (Bader and Winklhofer, 2020; Gomes de Oliveira et al., 2021).

4.3.4 Peripheral inflammation contribution

The gut microbiota is a major source of pro-inflammatory agents that stimulate immune cells of target organs. Compelling evidence indicates that alcohol increases gut permeability to macromolecules, a process known as "gut leakiness" that allows circulating cytokines to enter the brain and induce neuroimmune activation. Importantly, increased circulating cytokines in the periphery are associated with immune activation in the brain, which can persist for months after the peripheral immune response has subsided (reviewed in Gorky and Schwaber, 2016; Erickson et al., 2019; Leclercq et al., 2017).

Lipopolysaccharide (LPS) is one of the most common types of pathogen-associated molecular patterns (PAMPs), whose recognition is accomplished via TLR4 signaling, by releasing pro-inflammatory mediators in microglia, astrocytes, and neurons. Alcohol-induced neuroinflammation is increased after systemic LPS as a result of gut permeability. Excessive drinkers have higher levels of serum LPS and other markers of immune activation that correlate with the quantity of alcohol consumed, whereas levels of these markers decrease after abstinence. Thus, alcohol-induced peripheral immune dysregulation is a core characteristic of AUD and associated pathologies (Cannon et al., 2018).

4.3.5 TLR activation

Endogenous TLR agonists comprise an important signaling pathway between glia and neurons (Crews and Vetreno, 2016). These receptors activate signaling cascades that converge on NF-κB inducing cytokines release. Notably, the nuclear protein, high-mobility group box 1 (HMGB1, released by glia and neurons upon inflammasome activation) is a cytokine-like molecule that can directly activate TLRs, thus initiating neuroinflammation. It is proposed that ethanol induction of innate immune activation and TLR signaling contributes to alcohol-induced neurodegeneration (Pascual et al., 2015; Alfonso-Loeches et al., 2010b; Grantham et al., 2020). Twelve TLRs have been identified in mice, with TLR4/TLR2 required for ethanol-induced activation of microglia and subsequent release of IL-1b, TNF-a, MIP-1a, MIP-2, and IL-6 and other inflammatory mediators (NO and free radicals), which in turn, activate neuronal apoptosis (Crews et al., 2015, 2017).

In this line, Vetreno and Crews (2012) and Vetreno et al. (2013) demonstrated that ethanol increases the expression of HMGB1 and TLRs in the human brain, an effect that persists throughout ethanol-induced withdrawal. Moreover, other studies also reported increased expression of cytokines, HMGB1, and TLRs in the postmortem alcoholic brain (reviewed in Crews et al., 2017). Ethanol also activates other immune receptors, such as NLRP3 inflammasome in the astroglia and microglia of the cerebral cortex of chronic ethanol-treated mice, and in primary cultured astroglial and microglial cells. This activation induces the release of IL-1β, IL-18 and IL-33, which leads to cell death by pyroptosis. The NLRP3 inflammasome starts with TLR4 signaling that along with mitochondrial ROS production stimulates the inflammasome complex, caspase-1 activation and pro-inflammatory cytokine IL-1β release (Alfonso-Loeches et al., 2014; Guerri and Pascual, 2019).

4.3.6 Ethanol bioactivation, oxidative stress, mitochondrial dysfunction, and neuroinflammation

Brain ethanol oxidation to acetaldehyde along with the action of ALDH2 promotes ROS generation in the mitochondria, resulting in a redox imbalance due to the formation of excess NADH by altering the $NADH/NAD^+$ ratio, representing both steps mechanisms of the pathogenesis of alcohol neurotoxicity, not only in the brain but also in other metabolic-active organs. Thus, the consequences of cellular damage associated with oxidative stress and mitochondrial dysfunction cannot be dissociated from their

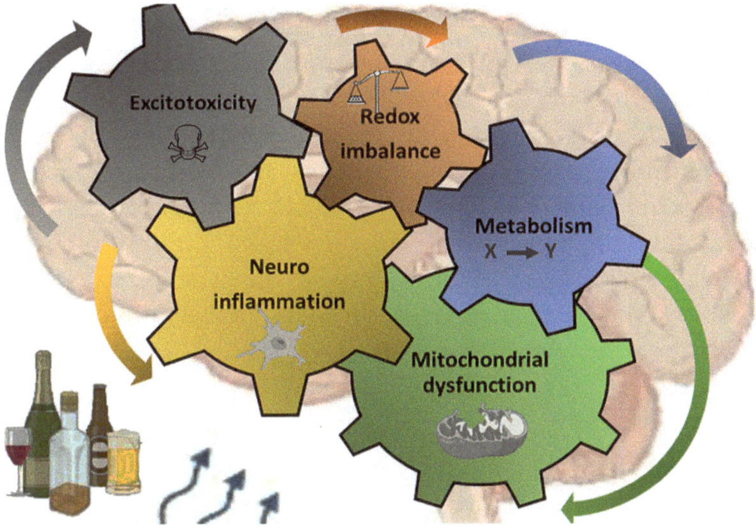

Fig. 1 Mechanisms that contribute to neuronal dysfunction and ethanol neurotoxicity and their interconnection.

interaction with the surrounding environment. Proinflammatory cytokines, such as TNF-α and IL-6 generate signals that affect mitochondrial ROS formation. Conversely, the injured cells generate reactive species that impair the capacity of the cells to respond appropriately to these cytokines. An understanding of the nature of these confluent and intertwined mechanisms in alcohol-induced neurotoxicity and its impact on cellular stress management will provide insights that may lead to new therapeutic approaches (see Fig. 1).

5. Animal models of ethanol-induced neurotoxicity

As mentioned, ethanol exerts potent neurotoxic effects. The mechanisms underlying these effects are varied, and include the induction of oxidative stress, the promotion of neuroinflammation, among others. This section will briefly review some of the ethanol administration patterns and models commonly used to induce neurotoxicity or neuronal degeneration in animal, mostly murine (rats, mice), animal models. Relevant findings derived from these methods, such as potential age-related differences in ethanol-induced neurotoxicity, will be discussed as well. The last segment of the section, in turn, will deal with newly established methods to assess the intriguing possibility that ethanol induces conditioned neuroinflammation.

5.1 The "Majchrowicz" model of binge ethanol intoxication

A widely used pre-clinical model for assessing ethanol-induced neurotoxicity is the "Majchrowicz" model (Collins et al., 1996; Corso et al., 1998; Crews et al., 2000; Tajuddin et al., 2014), developed almost 50 years ago with the main aim of inducing tolerance and withdrawal to ethanol (Majchrowicz, 1975). It is noteworthy mentioning that, at that moment, tolerance and withdrawal were considered the cardinal symptoms of alcohol-related disease. In the "Majchrowicz" preparation, which has been subjected to several variations, the animals are intubated into the stomach with high doses of ethanol (usually $\geq$4.0–5.0 g/kg) several times a day, for several days. This treatment induces extensive and widespread neural degeneration, across several brain structures. It also promotes a severe withdrawal response that can include, among other symptoms, audiogenic seizures. For instance, Liput et al. (2017) gave rats ethanol every 8 h for 1, 2, or 4 days. The 2-day treatment was associated with significant alterations in the endocannabinoid system at the hippocampus, one of the brain regions most likely to exhibit neurotoxicity after binge ethanol exposure.

Other studies with this model (Collins et al., 1996) reported substantial argyrophilia (i.e., high affinity for silver, an indicator of neural degeneration) in granular and pyramidal neurons cells of the dentate gyrus and entorhinal cortex, respectively. Liput et al. (2017), as other studies that employ this model, titrated the ethanol dose as a function of the behavioral monitoring of the animal. This is, an initial 5 g/kg dose was only repeated if the rat was deemed as "normal," otherwise lower doses were used (e.g., 4 or 2 g/kg if the rat was found to be hypoactive or ataxic). Confounds of this model are the reduction in water and food intake associated with the significant sedation induced by the ethanol treatment.

5.2 Age-related differences in ethanol-induced neurotoxicity

An interesting outcome found via derivatives of the "Majchrowicz" model is that adolescent rats seem to be more sensitive to ethanol-induced neural degeneration at several areas, including piriform and perirhinal cortices, when compared to adults (Crews et al., 2000). This finding is consistent with a study that employed an intermittent and lengthy ethanol exposure protocol (4 g/kg every 48 h, during a 20-day period). After such exposure, Broadwater et al. (2014) analyzed hippocampal neurogenesis, reporting reduced functionally in adolescents only. The protracted ethanol treatment in adolescents did not induce detrimental effects in the proliferation of neurons in the dentate gyrus but was associated with heightened levels of cleaved caspase-3 + IR. This finding indicates that lower neurogenesis after this ethanol exposure likely results from ethanol causing the death of new neurons within the dentate gyrus (Broadwater et al., 2014).

An intriguing study conducted in our lab (Fernández et al., 2019) also hinted at greater sensitivity to ethanol-induced neurotoxicity in adolescent vs. adults. Specifically, we exposed adolescent, adult or middle-age rats to 5 episodes of restraint (length: 2 h). The latter, which involves maintaining the animal in narrow tubes that prevent mobility, is considered a psychological stressor, mainly devoid of physical injury (Buynitsky and Mostofsky, 2009). Seventy hours after the last stressor the rats were given 2 injections of 2.5 g/kg ethanol (separated by 2 h) and were sacrificed 120 min after the last dosing. Neural degeneration was measured via the De Olmos Amino-Cupric-Silver technique (Bueno et al., 2003). Ethanol-induced, as expected, significant CA3 hippocampal degeneration, which was not affected by age or stress exposure. A significant new finding, however, was that adolescent, but not adult or middle-age rats, exhibited significant ethanol-induced argyrophilia at dorsal CA2, ventral CA3, and ventral DG. Perhaps more important, ethanol and stress had a significant interactive effect at ventral CA1 and CA2 (i.e., the cell degeneration found in animals exposed to both treatments was greater than in counterparts given one treatment or the other, but not both). This synergistic effect was restricted to adolescents (Fernández et al., 2019). Middle-age rats, in turn, were mostly spared of ethanol- or stress-induced neurotoxicity. Ventral CA1 yielded a similar degenerating profile (see representative photographs in Fig. 2), with the statistical analysis revealing a significant effect of stress in adolescent, but not in adult or middle-aged rats. These results are relevant due to the high prevalence of binge or high-intensity drinking (drinking 4–5 or 10 + standard units of alcohol in a single and usually brief occasion, respectively), which peaks at late adolescence and early adulthood.

5.3 Developmental neurotoxicity of alcohol

The findings described in the preceding section are not unexpected. Murine models of exposure to ethanol during the prenatal stage, or during the first weeks of life, have established that the developing brain is highly vulnerable to ethanol-induced neurotoxicity. When reviewing these studies is important to note the first 10 postnatal days (PD) in the rat or mice are characterized by a significant brain growth spurt, equivalent to that occurring during

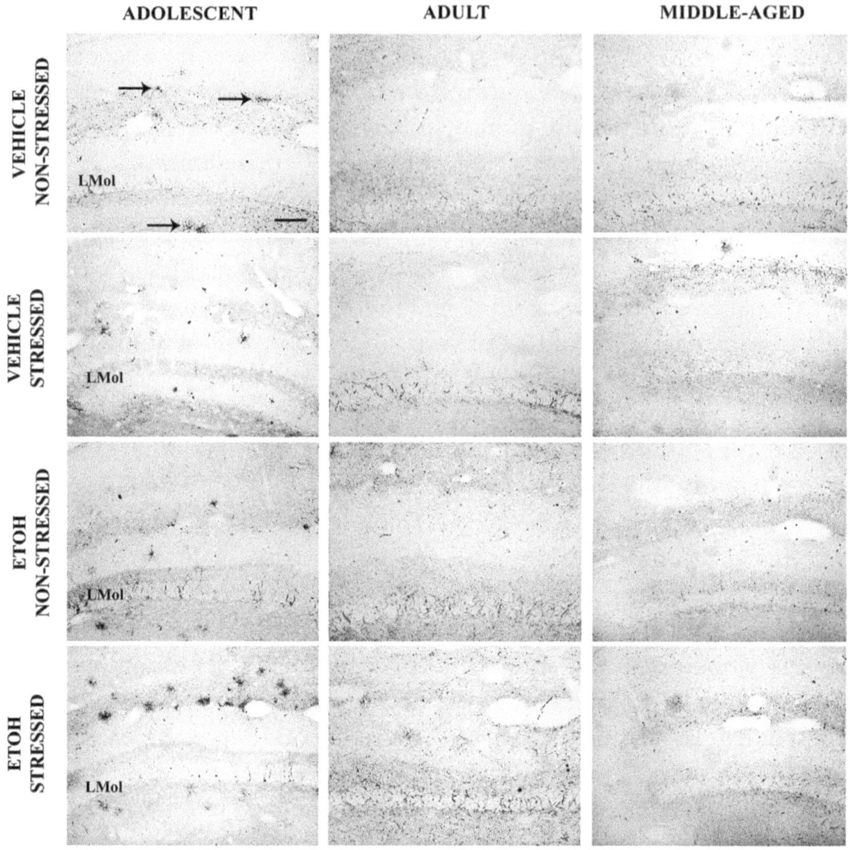

Fig. 2 Representative photographs depicting neuronal degeneration at hippocampus (CA1 section), as found by Fernández et al. (2019), via the use of the A-Cu-Ag technique. The rats, adolescent, adult or middle-aged Wistars, were restrained (stressed groups) once daily for 5 days or were left untreated in the home cage (non-stressed groups). Three days after the last stressor the rats received ethanol (2.5 g/kg i.p., separated by 2 h; vehicle: physiological saline) or vehicle, and were sacrificed 120 min after the last injection. The horizontal black line indicates a distance of 100 μm. The black arrows illustrate degenerating cells. The material is used with permission of the publisher.

the third trimester of gestation in humans (Kelly et al., 2009). Thus, murine models in which alcohol is given up to PD10 are actually modeling prenatal alcohol exposure in humans, which can give rise to a group of conditions grouped under the umbrella of fetal alcohol spectrum disorder (FASD).

Early work (Tomlinson et al., 1998) indicated that exposure to high-dose ethanol (6 g/kg, i.g.) at PD5—but not at PD10—was associated with lingering effects upon spatial learning, with these rats exhibiting significant deficits at PD61–64, when tested in the water Morris maze. Later work (Miki et al., 2004) linked this behavioral effect to significant loss of pyramidal cells in the hippocampus (CA1 and CA2 + CA3). In the latter study the rats had been exposed to relatively high blood alcohol concentrations (i.e., 432 ± 38 mg/dL), derived from exposure to inhaled ethanol at PD10–15 (3h/day). The high sensitivity of the developing brain to ethanol-induced neurotoxicity is also illustrated by a study (Balaszczuk et al., 2011) that gave preweanling rats just two administrations of subcutaneous alcohol (2.5 g/kgs.c., separated by 120 min) at PD7, 15 or just before weaning, at PD20. This brief ethanol exposure was enough to induce significant apoptotic neurodegeneration, as measured via the amino-cupric-silver technique, in the central extended amygdala on PD7 or 15, but not on PD20.

Quite common, among murine studies aiming at assessing neurotoxic effects of ethanol exposure during the first or second postnatal week, is to provide the drug in a milk solution via intragastric intubation (Farhadi et al., 2022). A recent study (Mohseni et al., 2021) provided rats such a mixture of ethanol (2.62 g/kg) and milk twice a day across PDs 2–10, and compared the effects of the treatment against untreated rats, rats intubated but not receiving any solution (i.e., a sham group) or against rats given ethanol plus apelin, an endogenous ligand for the G-protein-coupled receptor APJ (Tatemoto et al., 1998). Ethanol-exposed pups exhibited, when tested at PD36, significant deficits in spatial learning, and greater levels of CAT, GSH, and SOD dismutase in the hippocampus. Levels of the pro-inflammatory cytokine TNF-α were also increased after neonatal ethanol exposure. Interestingly, all these neurotoxic effects were inhibited or at least significantly alleviated by apelin, thus pinpointing to the potential translational value of this ligand.

It is important to note that effects similar to those found after neonatal ethanol exposure can be observed when the administration takes place during the adolescence of the rat. Oliveira et al. (2015) (also see Teixeira et al., 2014) gave female rats 6.5 g/kg/day from PD35 to PD90. At termination of this treatment the rats exhibited deficits in short-term, hippocampal-dependent, memory, an effect associated with significant neuronal loss in CA1, as well as reduced number of hippocampal astrocytes and microglia.

Altogether, these studies convincingly indicate that exposure to ethanol during infancy (or adolescence) can be associated with significant neurotoxicity that translates into neurobehavioral deficits. A drawback is their use of intensive and protracted ethanol exposure protocols that hardly mimic the sporadic pattern of alcohol exposure found in most human samples. Another study exposed adolescent or adult rats to ethanol (5.0 g/kg, i.p.) every other day over 20 days. This treatment, certainly intense yet not as much as that of Oliveira et al. (2015) or Teixeira et al. (2014), was not associated with immediate or short-term spatial working memory impairments. However, the adolescent but not the adult rats exhibited significant working memory deficits if challenged with ethanol.

Another widely used model for the assessment of ethanol-induced neurotoxicity focuses in the Purkinje cells in the cerebellum. These are very sensitive to ethanol-induced neurotoxicity. A study (Oliveira et al., 2014) exposed UChB rats (a strain characterized for exhibiting high levels of ethanol drinking) to ethanol from PD65 (at the transition from adolescence to adulthood) to PD120. The Purkinje cells in the cerebellum of these rats, which drank approximately 2 g/kg/day, exhibited greater levels of the pro-apoptotic protein caspase-3, which was consistent with signs of neural degeneration, including lipid droplets and densification of the nuclei. Significant alterations were also observed in Golgi and granular cells (e.g., pyknotic nuclei).

The damage caused by ethanol to Purkinje cells in the cerebellum is not uniform across development, it seems to be greater when the developmental insult takes place at PDs 4–6 of the rat, which we again emphasize is equivalent to the third trimester of gestation in humans (Pierce et al., 1999). As an example, Ramezani et al. (2012) gave male rat pups just two doses of 6 g/kg ethanol, at PDs 4 and 5, and sacrificed the rats 90 min after the last administration. This relatively brief treatment was associated with significant increases in the levels of thiobarbituric acid reactive substances (TBARS), a significant decrease in the activity of GPx and alterations in the layering and morphology of the cerebellar Purkinke cells. Perhaps even more impressive is that a single ethanol exposure between PD2 and PD8 yields significant neurotoxicity in these cells. More in detail, Idrus and Napper (2012) gave rats a single of administration of 4.5 g/kg ethanol, that could place any day between PD1 and 10. They found significant loss of Purkinje cell in cerebellar lobule III after PD2 exposure, whereas exposure in PD4 yielded a more general reduction across the structure.

The detrimental effects of ethanol on the cerebellum or hippocampus can also be observed when the drug is actually consumed by the mother during gestation. Children diagnosed with FASD exhibit reductions in the volume of several sections of the cerebellum (Mitoma et al., 2021), with alterations in the vermis (the section that connects the two hemispheres) probably being the most reliable sign of gestational ethanol exposure (Autti-Rämö et al., 2002). Animal models of prenatal ethanol exposure provided comparable outcomes (Sawant et al., 2013).

While discussing neurotoxic effects of prenatal ethanol exposure, it is noteworthy mentioning an elegant study by Uban et al. (2010). This research indicated that rats derived from dams that fed from an ethanol-containing liquid diet throughout gestation exhibited a significant reduction, as measured via injections of bromodeoxyuridine, in the number of new neurons or glial cells in the hippocampal dentate gyrus. This pattern is intriguingly similar to that reported by Broadwater et al. (2014) after adolescent (but not adult) exposure to ethanol and suggests that the "loss" of neural populations after prenatal ethanol exposure is in fact the consequence of the drug impacting the fate of newly produced neurons at hippocampus. It could be argued that prenatal or adolescent exposure to ethanol alters neural proliferation, yet none of these studies found a significant impact in this process.

The detrimental effects of prenatal ethanol exposure on neurogenesis were also highlighted by Choi et al. (2005). These researchers found a 2-fold increase in hippocampal neurogenesis in mice exposed to environmental enrichment, a protocol in which the mice are bred in larger-than-usual cages equipped with toys, ladders and the possibility of engaging in physical activity (Pautassi et al., 2017). Mice derived from dams exposed to ethanol exhibited a blunted neural response to this protocol. The study by Uban et al. (2010) employed a heavy exposure to ethanol, whereas Choi and co-workers exposed the pregnant dams to a two-bottle choice (ethanol vs. water) drinking protocol, usually associated with moderate (i.e., circa 4–5 g/kg/24 h) ethanol exposure.

5.4 Ethanol-induced conditioned neuroinflammation

So far, we reviewed preclinical studies interested in assessing ethanol-induced neurotoxicity, which typically administer several high doses of ethanol separated by short intervals. The studies that focus on ethanol-induced neuroinflammation generally administer lower ethanol doses, in a daily or every-other day fashion (e.g., in mice, 3 g/kg every other day for 2 weeks; Montesinos et al., 2016) that mimic patterns of binge drinking of young adults. These preclinical studies revealed microglial activation and brain expression of proinflammatory cytokines and chemokines (such as IL-1β, TNF-α and CCL-2) after acute or chronic alcohol intoxication, effects that persist up to 10 days in blood and liver (Qin et al., 2008a,b). Forced and protracted self-administration of ethanol in mice, leading to daily ingestion of approximately 12–14 g/kg/d, has been also shown to increase levels of pro-inflammatory cytokines (IL-1β, TNF-α), and gliosis processes in the cerebral cortex (Alfonso-Loeches et al., 2013).

These findings are consistent with clinical studies. One such study, conducted in blood samples from patients who fulfilled DSM-IV criteria for alcohol dependence, reported reduced activity of SOD, CAT and GPx, 14 days after the last ingestion of alcohol (Huang et al., 2009). These enzymes normally serve as deterrents against the oxidative stress caused by heightened intracellular levels of ROS which are in turn associated with chronic inflammation. A post-mortem analysis of brains of patients with a medical history of alcohol use disorder indicated heightened concentration of monocyte chemoattractant protein 1 (a chemokine involved in the migration and infiltration of macrophages; Deshmane et al., 2009) and microglia markers in hippocampus, amygdala, ventral tegmental area and substantia nigra. Moreover, in comorbidity with other factors (e.g., nutritional deficiency), the neuroinflammation caused by alcohol induces incapacitating conditions, such as Korsakoff's syndrome (Tuladhar and de Leeuw, 2019) or cerebellar degeneration and encephalopathies. More sporadic alcohol consumption, yet in the form of binge drinking, can also lead to cognitive impairment (Stavro et al., 2013).

Inflammation is a physiological response aimed at controlling the noxious stimulus triggering the process, yet its persistence is associated to significant damage and cell degeneration. Several studies have pointed to direct and indirect mechanisms of alcohol-induced neuroinflammation. In vitro studies have shown astroglial and microglial activation in response to ethanol through the TLR4, whereas deletion of the gene encoding the TLR4 decreases both alcohol-related microglial activation and neuronal death (Alfonso-Loeches et al., 2010a,b).

An intriguing possibility is that the neuroinflammation associated with ethanol exposure is perpetuated, perhaps months or years after the detoxification, via classical (i.e., Pavlovian) conditioning. Assessing this hypothesis has proven troublesome due to the inherent methodological difficulty of assessing the association between discrete conditioned stimuli (CS) and the complex signaling cascade leading to neuroinflammation. In two works (Gano et al., 2017, 2019), however, we have provided proof-of-concept evidence in favor of the hypothesis. The general strategy involved exposing rats to a novel and salient odor (CS, usually a lemon odor) during the course of the intoxication induced by moderate to high ethanol doses. The pairings lasted for 30 min and were conducted in a novel environment. The control group included animals given vehicle instead of ethanol or given unpaired exposure to ethanol and the putative CS (i.e., the odor). The harvesting of the brains was conducted 1 day after the last pairing, at variable times after re-exposure to the lemon CS. No significant alterations in brain gene expression of the proinflammatory cytokines IL-1, IL-6 or TNF-α was found, after one or two pairings. Yet, rats exposed to four pairings of 2.0 g/kg ethanol and lemon odor, and tested after receiving a priming dose of ethanol (i.e., 0.5 g/kg, which itself does not elicit cytokine expression), exhibited heightened IL-6 levels in hippocampus and amygdala than unpaired controls (Gano et al., 2017). These results suggest that the acquisition of ethanol-induced conditioned neuroinflammation requires several learning events, and that such response is observed by sensitizing a sub-threshold (in terms of the induction of proinflammatory cytokines) dose of ethanol.

In a subsequent study, however, we employed a taste-odor compound as CS and, probably due to the greater salience of this compound, were able to observe conditioned elevations in IL-6 at amygdala after a single ethanol-CS pairing (Gano et al., 2019). Interestingly, the magnitude of this effect was greater in adolescent than in young adult rats, again pinpointing the relatively greater sensitivity of immature subjects (infants, adolescents) to ethanol-induced neural degeneration. These findings, and particularly the experimental preparation that led to them, can be important for clinicians interested in developing interventions to extinguish learned associations that favor relapse into drinking.

5.5 Therapeutic interventions

5.5.1 Approved alcohol pharmacotherapy

Up to date, three medicines with modest efficacy are approved by the Food and Drug Administration (FDA) to treat AUD: disulfiram, naltrexone, and acamprosate. Disulfiram is considered a deterrent for alcohol consumption, that acts promoting toxic acetaldehyde build-up, whereas naltrexone is an opioid receptor antagonist that reduces the rewarding effects of alcohol. Acamprosate is a GABA analog, considered a NMDAR antagonist, used only during alcohol withdrawal. Last but not least, nalmefene, an analog to naltrexone, is recommended by the European Medicines Agency (EMA) for prescription to high alcohol drinkers (Maisel et al., 2013; Tizabi et al., 2021).

These medications will be presented in detail in the following section, while therapeutical approaches that are under study will be also mentioned in a mechanistic-based approach.

5.5.2 Metabolism

Disulfiram is considered a deterrent drug because inhibits ALDH, causing acetaldehyde to accumulate following the ingestion of alcohol. As both cytoplasmic and mitochondrial ALDH enzymes are inhibited, renewed enzyme activity depends on new protein synthesis, thus the effects can last several days after the last dose of disulfiram. The concomitant ingestion of alcohol and disulfiram produces severe aversive effects akin to those observed by Asians that have the inactive ALDH2*2 isoform, including facial flushing, nausea and vomiting, as well as alteration in blood pressure and headache. These symptoms discourage the patients to drink alcohol while on treatment with this drug (Barth and Malcolm, 2010).

Calcium carbamide (cyanamide) shares effects with disulfiram in that it interferes with the normal metabolism of alcohol by preventing the breakdown of acetaldehyde by ALDH inhibition. It was approved in Canada in 1959 but later withdrawn (Niederhofer et al., 2003). The first reports in animal models showed that systemic cyanamide depressed ethanol intake and locomotion (Sinclair et al., 1980; Spivak et al., 1987), while intra VTA cyanamide enhanced ethanol-induced locomotion (Martí-Prats et al., 2013), an effect supporting the notion that the central effects of acetaldehyde are reinforcing. In line with these results, we have reported opposing behavioral consequences of peripheral and central ALDH2 inhibition in rats. Liver ALDH2 activity was completely inhibited as a consequence of systemic cyanamide administration although both, ethanol intake and locomotion failed to decrease. In contrast, i.c.v. ALDH2 administration augmented ethanol intake and locomotion, in concomitance with brain ALDH2 activity and expression inhibition (Mattalloni et al., 2017, 2019).

5.5.3 Neuronal effects

As already reviewed in this chapter, alcohol affects multiple neurotransmitter systems: increases GABAergic activity, inhibits the glutamatergic neurotransmission through NMDA receptors, and increases DA release through endogenous opioids release. In line with this, compounds acting on GABAergic, opioid or glutamate systems, such as the GABAB receptor agonist baclofen, the opioid receptor antagonist naltrexone and the cystine precursor NAC that modulates glutamatergic neurotransmission, have received attention as possible treatment options for AUD. Studies in animal models reveal that each of these compounds reduces alcohol consumption and motivation. For instance, antagonism of mu opioid receptors decreased ethanol intake, ethanol-induced locomotion and conditioned place preference by ethanol. Clinical trials and individual responses after these compounds have been, however, less conclusive (Castrén et al., 2019). Additionally, the glutamate modulator acamprosate has been shown neuroprotective effects against glutamate-induced neurotoxicity and NMDAR hypersensitivity occurring during alcohol withdrawal, which may render neurons vulnerable to excitotoxicity and neuronal loss (Díaz-Soto and Calderín-Miranda, 2020).

5.6 Proposed alcohol pharmacotherapy

5.6.1 Metabolism

- ALDH

Alda-1 (N-[1, 3-benzodioxol-5-ylmethyl]-2, 6-dichlorobenzamide) is the first identified selective activator of ALDH2, acting as a molecular chaperone of the enzyme. ALDA-1 activates ALDH2 by increasing its affinity for NAD^+, thus decreasing the Km of ALDH2 for NAD^+, resulting in an increase in the catalytic efficiency. In addition, Alda-1 can bind to mutant ALDH2*1 and ALDH2*2, allowing the functioning of the critical active site residues of Cys302 and Glu268. Furthermore, Alda-1 can reduce accessibility of 4-HNE to Cys302, as well as the adjacent Cys301 and Cys303 residues, decreasing aldehyde-base adducts formation and reversing ALDH2 inhibition induced by 4-HNE. Interestingly, in addition to the direct combination of ALDH2, Alda-1 and Alda-44 (another ALDH2 activator) are able to activate ALDH2 through protein kinase C epsilon (εPKC)-mediated phosphorylation (Wang and He, 2018). Thus, ALDH2 can be activated by Alda-1, which represents a novel therapeutic target in cardiac and hepatic diseases, and can be employed in the treatment of AUD and neurodegenerative diseases that have been associated to ALDH malfunction (Chen et al., 2014; Rivera-Meza et al., 2019).

- CYP2E1

Diallyl sulfide (DAS) and other organosulfur antioxidants are constituents of garlic that have shown high efficiency in inhibiting CYP2E1. The ROS generated by the action of the CYP2E1 enzyme mediate the damage to DNA, proteins, and lipid membranes, ultimately causing organ damage. Given the known toxicity of DAS, it is desirable that novel analogs will be synthetized to treat AUD and other disease conditions, by targeting the CYP2E1 enzyme (Rao et al., 2015).

5.6.2 Mitochondrial functionality and gut dysbiosis

The development of new mitochondria-targeted antioxidants should be a priority in the development of new pharmacotherapeutic approaches. The antioxidant ability of glutathione, ascorbate or NAC should be complemented with other types of antioxidants that can scavenge radical reactions inside the mitochondria to be subsequently recharged by the ETC. At this respect, Mitochondria-targeted lipophilic ubiquinone (MitoQ) comprises coenzyme Q10 and has been shown to be protective against various oxidative stress-related processes associated with alcohol exposure.

In a recent study Hao et al. (2018) have demonstrated that MitoQ may speed up acetaldehyde detoxification by regulating ALDH2 post-translation modifications. The latter helped ameliorate oxidative stress and neuroinflammation in the alcohol-induced pathogenesis at the gut-liver axis, in a mouse model of alcohol liver disease. Although these results apply to peripheral acetaldehyde accumulation, the shared mechanisms between liver and brain alcohol toxicity allow to infer similar detoxicant pathways between both organs. In this respect, although brain ethanol metabolism is less oxidant than in the liver due to the absence of ADHs, it is noteworthy that there are other sources that affect the redox environment in the brain such as CYP2E1, and proinflammatory cytokines, among others. Last but not least, the contribution of the gut to neuroinflammation is a promising therapeutic target for patients suffering alcohol withdrawal symptoms, particularly given the safe and well-established methods of modulating the gut microbiome (Erickson et al., 2019; Gorky and Schwaber, 2016).

5.6.3 Metabolic consequences

Chronic ethanol consumption produces a progressive disruption of critical regulatory mechanisms that maintain metabolic homeostasis by decreasing the activity of oxidative phosphorylation, abnormally shaped mitochondria and other cellular morphological and enzymatic changes. Thus, in the event of alcohol withdrawal, multiple strategies should be implemented to mitigate the broad spectrum of adverse effects causes by drug discontinuation, in particular in the brain. This approach includes the correction of the energy deficiency primarily in neurons and astrocytes which can be achieved by providing a metabolite "cocktail" (butyrate, lactate, and β-hydroxybutyrate) for energy metabolism and support of the citric acid cycle. Other suitable nutrients include branched chain amino acids and branched short chain fatty acids, as these also rapidly cross the BBB. Importantly, supplementing with carnitine and/or acetyl-L-carnitine may also be helpful.

5.6.4 Glutamate homeostasis

A first strategy should aim to stabilize glutamate distribution and metabolism in the brain. In this regard, the treatment with NAC has been reported to enhance the expression of xCT, increasing the concentrations of cysteine and cystine, and indirectly activating mGlu2 receptors which altogether helps rebalance glutamate homeostasis and glutamatergic neuronal/astrocyte interactions. In fact, studies in animals demonstrated that NAC prevents behavioral consequences of ethanol withdrawal and alcohol-related neuroinflammation (Schneider et al., 2017, 2015). Moreover, NAC treatment blocks the development of ethanol-induced behavioral sensitization, a behavioral model of the transitions from a non-addicted to an addicted-state, and related neuroplasticity (Morais-Silva et al., 2016) whereas the xCT expression is related to individual susceptibility to ethanol-induced conditioned place preference (Amaral et al., 2021). Interestingly, ceftriaxone, a β-lactam antibiotic, reduced alcohol intake in alcohol-preferring rats (Alhaddad et al., 2014; Rao and Sari, 2014; Stennett et al., 2017).

The pharmacological compensation for the increased sensitivity of the glutamatergic neural system should be also addressed. As mentioned, chronic ethanol consumption suppresses glutamatergic activity, which results in an increase in the excitatory NMDAR and AMPAR and a decrease in inhibitory mGluR. The contribution of enhanced glutamatergic sensitivity and neurotoxicity-derived from excessive NMDAR activity– to ethanol withdrawal symptoms can be suppressed by antagonists of excitatory glutamate receptors, such as acamprosate, memantine, or MK-801 (Wilson and Matschinsky, 2020).

5.6.5 Neuroinflammation

Up to date, the opioid receptor antagonist naltrexone is the only drug—among those approved by the FDA to treat AUD– with potential to target the neuroimmune system. This is because naltrexone has been proposed to exert an antagonistic activity upon TLR4 receptors (Guglielmo et al., 2015). Nalmefene, like naltrexone, is a μ-receptor antagonist and weak κ-agonist currently only approved in Europe to reduce heavy drinking. Its effects on TLR4, however, are yet to be described. Other immunomodulators such as PPAR agonists or ibudilast are under study (Erickson et al., 2019). Additionally, it has been shown that minocycline, a tetracycline antibiotic, decreased alcohol intake (Agrawal et al., 2011), reduced alcohol induced place preference, diminished the rewarding effect of alcohol and attenuated alcohol relapse (Gajbhiye et al., 2017). Other, off-label medications, such as baclofen, gabapentin, ondansetron and topiramate are commonly prescribed for the treatment of AUD. Sodium oxybate is approved in Italy and Austria only and baclofen received a "temporary recommendation for use" in France. Encouraging results have been reported for topiramate, gabapentin and also varenicline (Antonelli et al., 2018; Castrén et al., 2019).

6. Conclusions and future considerations

The relationship between alcohol-induced brain damage and the development of AUD requires additional experimentation and the execution of carefully controlled clinical trials. The unravel of the intertwined molecular and cellular effects is necessary to propose new pharmacological treatments. This would certainly have a positive impact on reducing alcohol seeking, craving and on the recovery from AUD.

Thereby, due to multifactorial nature of AUD, is unlikely that a single drug will single-handedly mitigate the deleterious effects of chronic alcohol consumption and withdrawal symptoms. Thus, an effective treatment is likely to require simultaneous approaches. Small attempts such as NAC administration in concomitance with naltrexone has been shown to augment the efficacy of naltrexone therapy aimed to reduce alcohol dependence. More efforts should be posited to successfully prevent the spiral of neurotoxicant bioactivation, redox imbalance, mitochondrial dysfunction and inflammasome that characterize AUD and related pathologies.

Acknowledgments

Supported in part by grants from the Secretarıa de Ciencia y Tecnologia (SeCyT-UNC), and Fondo para la Investigación Científica y Tecnológica (FONCyT) from Argentina, awarded to MBV (PICT-2017-0874 and PICT-2019-02444) and RMP (PICT-2019-00180 and PICT 2018-00597).

References

Abbott, N.J., Rönnbäck, L., Hansson, E., 2006. Astrocyte–endothelial interactions at the blood–brain barrier. Nat. Rev. Neurosci. 7 (1), 41–53.

Adermark, L., Bowers, M.S., 2016. Disentangling the role of astrocytes in alcohol use disorder. Alcohol. Clin. Exp. Res. 40 (9), 1802–1816.

Aebi, H., 1984. [13] Catalase in vitro. In: Methods in Enzymology. vol. 105. Academic Press, pp. 121–126.

Agrawal, R., Hewetson, A., George, C.M., Syapin, P.J., Bergeson, S., 2011. Minocycline reduces ethanol drinking. Brain Behav. Immun. 25, S165–S169.

Alfonso-Loeches, S., Pascual-Lucas, M., Blanco, A.M., Sanchez-Vera, I., Guerri, C., 2010a. Pivotal role of TLR4 receptors in alcohol-induced neuroinflammation and brain damage. J. Neurosci. 30, 8285–8295.

Alfonso-Loeches, S., Pascual-Lucas, M., Blanco, A.M., Sanchez-Vera, I., Guerri, C., 2010b. Pivotal role of TLR4 receptors in alcohol-induced neuroinflammation and brain damage. J. Neurosci. 30 (24), 8285–8295.
Alfonso-Loeches, S., Pascual, M., Guerri, C., 2013. Gender differences in alcohol-induced neurotoxicity and brain damage. Toxicology 311 (1–2), 27–34.
Alfonso-Loeches, S., Ureña-Peralta, J.R., Morillo-Bargues, M.J., Guerri, C., 2014. Role of mitochondria ROS generation in ethanol-induced NLRP3 inflammasome activation and cell death in astroglial cells. Front. Cell. Neurosci. 8, 216.
Alhaddad, H., Das, S.C., Sari, Y., 2014. Effects of ceftriaxone on ethanol intake: a possible role for xCT and GLT-1 isoforms modulation of glutamate levels in P rats. Psychopharmacology 231 (20), 4049–4057.
Amaral, V.C.S., Morais-Silva, G., Laverde, C.F., Marin, M.T., 2021. Susceptibility to extinction and reinstatement of ethanol-induced conditioned place preference is related to differences in astrocyte cystine-glutamate antiporter content. Neurosci. Res. 170, 245–254.
Antonelli, M., Ferrulli, A., Sestito, L., Vassallo, G.A., Tarli, C., Mosoni, C., Rando, M.M., Mirijello, A., Gasbarrini, A., Addolorato, G., 2018. Alcohol addiction-the safety of available approved treatment options. Expert Opin. Drug Saf. 17 (2), 169–177.
Autti-Rämö, I., Autti, T., Korkman, M., Kettunen, S., Salonen, O., Valanne, L., 2002. MRI findings in children with school problems who had been exposed prenatally to alcohol. Dev. Med. Child Neurol. 44 (2), 98–106.
Bader, V., Winklhofer, K.F., 2020, March. Mitochondria at the interface between neurodegeneration and neuroinflammation. In: Seminars in Cell and Developmental Biology. vol. 99. Academic Press, pp. 163–171.
Baker, D.A., Xi, Z.X., Shen, H., Swanson, C.J., Kalivas, P.W., 2002. The origin and neuronal function of in vivo nonsynaptic glutamate. J. Neurosci. 22 (20), 9134–9141.
Balaszczuk, V., Bender, C., Pereno, G.L., Beltramino, C.A., 2011. Alcohol-induced neuronal death in central extended amygdala and pyriform cortex during the postnatal period of the rat. Int. J. Dev. Neurosci. 29 (7), 733–742.
Banerjee, K., Munshi, S., Sen, O., Pramanik, V., Roy Mukherjee, T., Chakrabarti, S., 2014. Dopamine cytotoxicity involves both oxidative and nonoxidative pathways in SH-SY5Y cells: potential role of alpha-synuclein overexpression and proteasomal inhibition in the etiopathogenesis of Parkinson's disease. Parkinson's Dis. 2014, 878935. https://doi.org/10.1155/2014/878935.
Barth, K.S., Malcolm, R.J., 2010. Disulfiram: an old therapeutic with new applications. CNS Neurol. Disord. Drug Targets 9 (1), 5–12 (formerly current drug targets-CNS & Neurological Disorders).
Bassareo, V., Frau, R., Maccioni, R., Caboni, P., Manis, C., Peana, A.T., Migheli, R., Porru, S., Acquas, E., 2021. Ethanol-dependent synthesis of salsolinol in the posterior ventral tegmental area as key mechanism of ethanol's action on mesolimbic dopamine. Front. Neurosci 15, p.675061.
Belarbi, K., Cuvelier, E., Destée, A., Gressier, B., Chartier-Harlin, M.C., 2017. NADPH oxidases in Parkinson's disease: a systematic review. Mol. Neurodegener. 12 (1), 1–18.
Berríos-Cárcamo, P., Quintanilla, M.E., Herrera-Marschitz, M., Vasiliou, V., Zapata-Torres, G., Rivera-Meza, M., 2017. Racemic salsolinol and its enantiomers act as agonists of the μ-opioid receptor by activating the Gi protein-adenylate cyclase pathway. Front. Behav. Neurosci. 10, 253.
Berríos-Cárcamo, P., Quezada, M., Quintanilla, M.E., Morales, P., Ezquer, M., Herrera-Marschitz, M., Israel, Y., Ezquer, F., 2020. Oxidative stress and neuroinflammation as a pivot in drug abuse. A focus on the therapeutic potential of antioxidant and anti-inflammatory agents and biomolecules. Antioxidants 9 (9), 830.

Beynon, S.B., Walker, F.R., 2012. Microglial activation in the injured and healthy brain: what are we really talking about? Practical and theoretical issues associated with the measurement of changes in microglial morphology. Neuroscience 225, 162–171.
Blednov, Y.A., Benavidez, J.M., Black, M., Ferguson, L.B., Schoenhard, G.L., Goate, A.M., Edenberg, H.J., Wetherill, L., Hesselbrock, V., Foroud, T., Adron Harris, R., 2015. Peroxisome proliferator-activated receptors α and γ are linked with alcohol consumption in mice and withdrawal and dependence in humans. Alcohol. Clin. Exp. Res. 39 (1), 136–145.
Boyadjieva, N.I., Sarkar, D.K., 2013. Microglia play a role in ethanol-induced oxidative stress and apoptosis in developing hypothalamic neurons. Alcohol. Clin. Exp. Res. 37 (2), 252–262.
Bridges, R., Lutgen, V., Lobner, D., Baker, D.A., 2012. Thinking outside the cleft to understand synaptic activity: contribution of the cystine-glutamate antiporter (system xc −) to normal and pathological glutamatergic signaling. Pharmacol. Rev. 64 (3), 780–802.
Broadwater, M.A., Liu, W., Crews, F.T., Spear, L.P., 2014. Persistent loss of hippocampal neurogenesis and increased cell death following adolescent, but not adult, chronic ethanol exposure. Dev. Neurosci. 36 (3–4), 297–305.
Bueno, A., De Olmos, S., Heimer, L., De Olmos, J., 2003. NMDA-antagonist MK-801-induced neuronal degeneration in Wistar rat brain detected by the amino-cupric-silver method. Exp. Toxicol. Pathol. 54 (4), 319–334.
Buynitsky, T., Mostofsky, D.I., 2009. Restraint stress in biobehavioral research: recent developments. Neurosci. Biobehav. Rev. 33 (7), 1089–1098.
Cannon, A.R., Hammer, A.M., Choudhry, M.A., 2018. Alcohol, inflammation, and depression: the gut-brain axis. In: Inflammation and Immunity in Depression. Academic Press, pp. 509–524.
Carocci, A., Catalano, A., Sinicropi, M.S., Genchi, G., 2018. Oxidative stress and neurodegeneration: the involvement of iron. Biometals 31 (5), 715–735.
Castrén, S., Mäkelä, N., Alho, H., 2019. Selecting an appropriate alcohol pharmacotherapy: review of recent findings. Curr. Opin. Psychiatry 32 (4), 266–274.
Cederbaum, A.I., Lu, Y., Wu, D., 2009. Role of oxidative stress in alcohol-induced liver injury. Arch. Toxicol. 83 (6), 519–548.
Chen, C.H., Ferreira, J.C.B., Gross, E.R., Mochly-Rosen, D., 2014. Targeting aldehyde dehydrogenase 2: new therapeutic opportunities. Physiol. Rev. 94 (1), 1–34.
Choi, I.Y., Allan, A.M., Cunningham, L.A., 2005. Moderate fetal alcohol exposure impairs the neurogenic response to an enriched environment in adult mice. Alcohol. Clin. Exp. Res. 29 (11), 2053–2062.
Collins, M.A., Neafsey, E.J., 2012. Neuroinflammatory pathways in binge alcohol-induced neuronal degeneration: oxidative stress cascade involving aquaporin, brain edema, and phospholipase A2 activation. Neurotox. Res. 21 (1), 70–78.
Collins, M.A., Corso, T.D., Neafsey, E.J., 1996. Neuronal degeneration in rat cerebrocortical and olfactory regions during subchronic "binge" intoxication with ethanol: possible explanation for olfactory deficits in alcoholics. Alcohol. Clin. Exp. Res. 20 (2), 284–292.
Comporti, M., Signorini, C., Leoncini, S., Gardi, C., Ciccoli, L., Giardini, A., Vecchio, D., Arezzini, B., 2010. Ethanol-induced oxidative stress: basic knowledge. Genes Nutr. 5 (2), 101–109.
Corso, T.D., Mostafa, H.M., Collins, M.A., Neafsey, E.J., 1998. Brain neuronal degeneration caused by episodic alcohol intoxication in rats: effects of nimodipine, 6,7-dinitro-quinoxaline-2, 3-dione, and MK-801. Alcohol. Clin. Exp. Res. 22 (1), 217–224.
Crews, F.T., Vetreno, R.P., 2016. Mechanisms of neuroimmune gene induction in alcoholism. Psychopharmacology 233 (9), 1543–1557.

Crews, F.T., Braun, C.J., Hoplight, B., Switzer III, R.C., Knapp, D.J., 2000. Binge ethanol consumption causes differential brain damage in young adolescent rats compared with adult rats. Alcohol. Clin. Exp. Res. 24 (11), 1712–1723.
Crews, F.T., Zou, J., Qin, L., 2011. Induction of innate immune genes in brain create the neurobiology of addiction. Brain Behav. Immun. 25, S4–S12.
Crews, F.T., Sarkar, D.K., Qin, L., Zou, J., Boyadjieva, N., Vetreno, R.P., 2015. Neuroimmune function and the consequences of alcohol exposure. Alcohol Res. 37 (2), 331.
Crews, F.T., Lawrimore, C.J., Walter, T.J., Coleman Jr., L.G., 2017. The role of neuroimmune signaling in alcoholism. Neuropharmacology 122, 56–73.
Crews, F.T., Zou, J., Coleman Jr., L.G., 2021. Extracellular microvesicles promote microglia-mediated pro-inflammatory responses to ethanol. J. Neurosci. Res. 99 (8), 1940–1956.
Cui, J., Liu, Y., Chang, X., Gou, W., Zhou, X., Liu, Z., Li, Z., Wu, Y., Zuo, D., 2019. Acetaldehyde induces neurotoxicity in vitro via oxidative stress-and Ca2 + imbalance-mediated endoplasmic reticulum stress. Oxidative Med. Cell. Longev. 2019, 2593742.
Das, S.K., Vasudevan, D.M., 2007. Alcohol-induced oxidative stress. Life Sci. 81 (3), 177–187.
Dashty, M., 2013. A quick look at biochemistry: carbohydrate metabolism. Clin. Biochem. 46 (15), 1339–1352.
de Oliveira, L.G., Angelo, Y.D.S., Iglesias, A.H., Peron, J.P.S., 2021. Unraveling the link between mitochondrial dynamics and neuroinflammation. Front. Immunol. 12, 752.
Deitrich, R., Zimatkin, S., Pronko, S., 2006. Oxidation of ethanol in the brain and its consequences. Alcohol Res. Health 29 (4), 266.
Deshmane, S.L., Kremlev, S., Amini, S., Sawaya, B.E., 2009. Monocyte chemoattractant protein-1 (MCP-1): an overview. J. Interferon Cytokine Res. 29, 313–326.
Di Filippo, M., Chiasserini, D., Tozzi, A., Picconi, B., Calabresi, P., 2010. Mitochondria and the link between neuroinflammation and neurodegeneration. J. Alzheimers Dis. 20 (s2), S369–S379.
Díaz-Soto, M.T., Calderín-Miranda, J.M., 2020. Alcohol withdrawal syndrome: result of oxidative stress and neuronal imbalance. State of the art. Rev. Biomed. 31 (2), 95–107.
Dixon, S.J., Lemberg, K.M., Lamprecht, M.R., Skouta, R., Zaitsev, E.M., Gleason, C.E., Patel, D.N., Bauer, A.J., Cantley, A.M., Yang, W.S., Morrison III, B., 2012. Ferroptosis: an iron-dependent form of nonapoptotic cell death. Cell 149 (5), 1060–1072.
Doorn, J.A., Hurley, T.D., Petersen, D.R., 2006. Inhibition of human mitochondrial aldehyde dehydrogenase by 4-hydroxynon-2-enal and 4-oxonon-2-enal. Chem. Res. Toxicol. 19 (1), 102–110.
Edenberg, H.J., McClintick, J.N., 2018. Alcohol dehydrogenases, aldehyde dehydrogenases, and alcohol use disorders: a critical review. Alcohol. Clin. Exp. Res. 42 (12), 2281–2297.
Erickson, E.K., Grantham, E.K., Warden, A.S., Harris, R.A., 2019. Neuroimmune signaling in alcohol use disorder. Pharmacol. Biochem. Behav. 177, 34–60.
Ezquer, F., Quintanilla, M.E., Morales, P., Ezquer, M., Lespay-Rebolledo, C., Herrera-Marschitz, M., Israel, Y., 2019a. Activated mesenchymal stem cell administration inhibits chronic alcohol drinking and suppresses relapse-like drinking in high-alcohol drinker rats. Addict. Biol. 24 (1), 17–27.
Ezquer, F., Quintanilla, M.E., Morales, P., Santapau, D., Ezquer, M., Kogan, M.J., Salas-Huenuleo, E., Herrera-Marschitz, M., Israel, Y., 2019b. Intranasal delivery of mesenchymal stem cell-derived exosomes reduces oxidative stress and markedly inhibits ethanol consumption and post-deprivation relapse drinking. Addict. Biol. 24 (5), 994–1007.

Farhadi, L., Hojati, V., Khaksari, M., Vaezi, G., 2022. Neuroprotective effects of Crocin against ethanol neurotoxicity in the animal model of fetal alcohol Spectrum disorders. Neurochem. Res., 1–11.
Farina, M., Avila, D.S., Da Rocha, Aschner, M., 2013. Metals, oxidative stress and neurodegeneration: a focus on iron, manganese and mercury. Neurochem. Int. 62 (5), 575–594.
Ferguson, L.B., Most, D., Blednov, Y.A., Harris, R.A., 2014. PPAR agonists regulate brain gene expression: relationship to their effects on ethanol consumption. Neuropharmacology 86, 397–407.
Fernandes, L.M.P., de Andrade Jr, E.F., Monteiro, M.C., Cartágenes, S.C., Lima, R.R., Prediger, R.D., Maia, C.S.F., 2017. Ethanol: neurotoxicity and brain disorders. In: Addictive Substances and Neurological Disease. Academic Press, pp. 201–215.
Fernández, M.S., de Olmos, S., Nizhnikov, M.E., Pautassi, R.M., 2019. Restraint stress exacerbates cell degeneration induced by acute binge ethanol in the adolescent, but not in the adult or middle-aged, brain. Behav. Brain Res. 364, 317–327.
Fernández-Rodríguez, S., Cano-Cebrián, M.J., Rius-Pérez, S., Pérez, S., Guerri, C., Granero, L., Zornoza, T., Polache, A., 2022. Different brain oxidative and neuroinflammation status in rats during prolonged abstinence depending on their ethanol relapse-like drinking behavior: effects of ethanol reintroduction. Drug Alcohol Depend. 232, 109284.
French, S.W., 2016. Chronic alcohol binging injures the liver and other organs by reducing NAD+ levels required for sirtuin's deacetylase activity. Exp. Mol. Pathol. 100 (2), 303–306.
Gajbhiye, S.V., Tripathi, R.K., Salve, B., Petare, A., Potey, A.V., 2017. Evaluation of effect of minocycline on rewarding potential and alcohol relapse in place preference model in mice. Neurosci. Lett. 649, 28–33.
Gano, A., Pautassi, R.M., Doremus-Fitzwater, T.L., Deak, T., 2017. Conditioned effects of ethanol on the immune system. Exp. Biol. Med. (Maywood) 242, 718–730.
Gano, A., Pautassi, R.M., Doremus-Fitzwater, T.L., Barney, T.M., Vore, A.S., Deak, T., 2019. Conditioning the neuroimmune response to ethanol using taste and environmental cues in adolescent and adult rats. Exp. Biol. Med. 244 (5), 362–371.
Goldstein, D.B., 1986. Effect of alcohol on cellular membranes. Ann. Emerg. Med. 15 (9), 1013–1018.
Goldstein, D.S., 2020. The catecholaldehyde hypothesis: where MAO fits in. J. Neural Transm. 127 (2), 169–177.
Gorky, J., Schwaber, J., 2016. The role of the gut–brain axis in alcohol use disorders. Prog. Neuro-Psychopharmacol. Biol. Psychiatry 65, 234–241.
Grantham, E.K., Warden, A.S., McCarthy, G.S., DaCosta, A., Mason, S., Blednov, Y., Mayfield, R.D., Harris, R.A., 2020. Role of toll-like receptor 7 (TLR7) in voluntary alcohol consumption. Brain Behav. Immun. 89, 423–432.
Guerri, C., Pascual, M., 2019. Role of neuroinflammation in ethanol neurotoxicity. In: Advances in Neurotoxicology. vol. 3. Academic Press, pp. 259–294.
Guglielmo, R., Martinotti, G., Quatrale, M., Ioime, L., Kadilli, I., Di Nicola, M., Janiri, L., 2015. Topiramate in alcohol use disorders: review and update. CNS Drugs 29 (5), 383–395.
Haile, C.N., Kosten, T.A., 2017. The peroxisome proliferator-activated receptor alpha agonist fenofibrate attenuates alcohol self-administration in rats. Neuropharmacology 116, 364–370.
Halliwell, B., 1992. Reactive oxygen species and the central nervous system. J. Neurochem. 59 (5), 1609–1623.
Hao, L., Sun, Q., Zhong, W., Zhang, W., Sun, X., Zhou, Z., 2018. Mitochondria-targeted ubiquinone (MitoQ) enhances acetaldehyde clearance by reversing alcohol-induced posttranslational modification of aldehyde dehydrogenase 2: a molecular mechanism of protection against alcoholic liver disease. Redox Biol. 14, 626–636.

Haorah, J., Ramirez, S.H., Floreani, N., Gorantla, S., Morsey, B., Persidsky, Y., 2008. Mechanism of alcohol-induced oxidative stress and neuronal injury. Free Radic. Biol. Med. 45 (11), 1542–1550.
Hartman, J.H., Miller, G.P., Meyer, J.N., 2017. Toxicological implications of mitochondrial localization of CYP2E1. Toxicology Research 6 (3), 273–289.
Heit, C., Dong, H., Chen, Y., Thompson, D.C., Deitrich, R.A., Vasiliou, V.K., 2013. The role of CYP2E1 in alcohol metabolism and sensitivity in the central nervous system. In: Cytochrome P 450 2E1: Its Role in Disease and Drug Metabolism. Springer, Dordrecht, pp. 235–247.
Henriques, J.F., Portugal, C.C., Canedo, T., Relvas, J.B., Summavielle, T., Socodato, R., 2018. Microglia and alcohol meet at the crossroads: microglia as critical modulators of alcohol neurotoxicity. Toxicol. Lett. 283, 21–31.
Huang, M.-C., Chen, C.-H., Peng, F.-C., Tang, S.-H., Chen, C.-C., 2009. Alterations in oxidative stress status during early alcohol withdrawal in alcoholic patients. J. Formos. Med. Assoc. 108, 560–569.
Ibáñez, F., Ureña-Peralta, J.R., Costa-Alba, P., Torres, J.L., Laso, F.J., Marcos, M., Guerri, C., Pascual, M., 2020. Circulating micrornas in extracellular vesicles as potential biomarkers of alcohol-induced neuroinflammation in adolescence: gender differences. Int. J. Mol. Sci. 21 (18), 6730.
Idrus, N.M., Napper, R.M.A., 2012. Acute and long-term Purkinje cell loss following a single ethanol binge during the early third trimester equivalent in the rat. Alcohol. Clin. Exp. Res. 36, 1365–1373.
Israel, Y., Rivera-Meza, M., Karahanian, E., Quintanilla, M.E., Tampier, L., Morales, P., Herrera-Marschitz, M., 2013. Gene specific modifications unravel ethanol and acetaldehyde actions. Front. Behav. Neurosci. 7, 80.
Israel, Y., Quintanilla, M.E., Karahanian, E., Rivera-Meza, M., Herrera-Marschitz, M., 2015. The "first hit" toward alcohol reinforcement: role of ethanol metabolites. Alcohol. Clin. Exp. Res. 39 (5), 776–786.
Israel, Y., Quintanilla, M.E., Ezquer, F., Morales, P., Santapau, D., Berríos-Cárcamo, P., Ezquer, M., Olivares, B., Herrera-Marschitz, M., 2021. Aspirin and N-acetylcysteine co-administration markedly inhibit chronic ethanol intake and block relapse binge drinking: role of neuroinflammation-oxidative stress self-perpetuation. Addict. Biol. 26 (1), e12853.
Karahanian, E., Quintanilla, M.E., Fernandez, K., Israel, Y., 2014. Fenofibrate–a lipid-lowering drug–reduces voluntary alcohol drinking in rats. Alcohol 48 (7), 665–670.
Kelly, S.J., Goodlett, C.R., Hannigan, J.H., 2009. Animal models of fetal alcohol spectrum disorders: impact of the social environment. Dev. Disabil. Res. Rev. 15, 200–208.
Khanna, A., Acharjee, P., Acharjee, A., Trigun, S.K., 2019. Mitochondrial SIRT3 and neurodegenerative brain disorders. J. Chem. Neuroanat. 95, 43–53.
Lachenmeier, D.W., Kanteres, F., Rehm, J., 2009. Carcinogenicity of acetaldehyde in alcoholic beverages: risk assessment outside ethanol metabolism. Addiction 104 (4), 533–550.
Lebourgeois, S., González-Marín, M.C., Jeanblanc, J., Naassila, M., Vilpoux, C., 2018. Effect of N-acetylcysteine on motivation, seeking and relapse to ethanol self-administration. Addict. Biol. 23 (2), 643–652.
Lebourgeois, S., González-Marín, M.C., Antol, J., Naassila, M., Vilpoux, C., 2019. Evaluation of N-acetylcysteine on ethanol self-administration in ethanol-dependent rats. Neuropharmacology 150, 112–120.
Leclercq, S., de Timary, P., Delzenne, N.M., Stärkel, P., 2017. The link between inflammation, bugs, the intestine and the brain in alcohol dependence. Transl. Psychiatry 7 (2), e1048.
Lieber, C.S., 2005. Metabolism of alcohol. Clin. Liver Dis. 9 (1), 1–35.

Liput, D.J., Pauly, J.R., Stinchcomb, A.L., Nixon, K., 2017. Binge alcohol exposure transiently changes the endocannabinoid system: a potential target to prevent alcohol-induced neurodegeneration. Brain Sci. 7 (12), 158.

Liu, C.Y., Wang, M., Yu, H.M., Han, F.X., Wu, Q.S., Cai, X.J., Kurihara, H., Chen, Y.X., Li, Y.F., He, R.R., 2020. Ferroptosis is involved in alcohol-induced cell death in vivo and in vitro. Biosci. Biotechnol. Biochem. 84 (8), 1621–1628.

Maisel, N.C., Blodgett, J.C., Wilbourne, P.L., Humphreys, K., Finney, J.W., 2013. Meta-analysis of naltrexone and acamprosate for treating alcohol use disorders: when are these medications most helpful? Addiction 108 (2), 275–293.

Majchrowicz, E., 1975. Induction of physical dependence upon ethanol and the associated behavioral changes in rats. Psychopharmacologia 43, 245–254.

Manzo-Avalos, S., Saavedra-Molina, A., 2010. Cellular and mitochondrial effects of alcohol consumption. Int. J. Environ. Res. Public Health 7 (12), 4281–4304.

Martí-Prats, L., Sánchez-Catalán, M.J., Orrico, A., Zornoza, T., Polache, A., Granero, L., 2013. Opposite motor responses elicited by ethanol in the posterior VTA: the role of acetaldehyde and the non-metabolized fraction of ethanol. Neuropharmacology 72, 204–214.

Mattalloni, M.S., De Giovanni, L.N., Molina, J.C., Cancela, L.M., Virgolini, M.B., 2013. Participation of catalase in voluntary ethanol consumption in perinatally low-level lead-exposed rats. Alcohol. Clin. Exp. Res. 37 (10), 1632–1642.

Mattalloni, M.S., Deza-Ponzio, R., Albrecht, P.A., Cancela, L.M., Virgolini, M.B., 2017. Developmental lead exposure induces opposite effects on ethanol intake and locomotion in response to central vs. systemic cyanamide administration. Alcohol 58, 1–11.

Mattalloni, M.S., Deza-Ponzio, R., Albrecht, P.A., Fernandez-Hubeid, L.E., Cancela, L.M., Virgolini, M.B., 2019. Brain ethanol-metabolizing enzymes are differentially expressed in lead-exposed animals after voluntary ethanol consumption: pharmacological approaches. Neurotoxicology 75, 174–185.

Mayfield, J., Ferguson, L., Harris, R.A., 2013. Neuroimmune signaling: a key component of alcohol abuse. Curr. Opin. Neurobiol. 23 (4), 513–520.

Meyer, J.N., Leuthner, T.C., Luz, A.L., 2017. Mitochondrial fusion, fission, and mitochondrial toxicity. Toxicology 391, 42–53.

Meredith, L.R., Burnette, E.M., Grodin, E.N., Irwin, M.R., Ray, L.A., 2021. Immune treatments for alcohol use disorder: a translational framework. Brain Behav. Immun. 97, 349–364.

Meyer, J.N., Hartman, J.H., Mello, D.F., 2018. Mitochondrial toxicity. Toxicol. Sci. 162 (1), 15–23.

Miki, T., Harris, S.J., Wilce, P.A., Takeuchi, Y., Bedi, K.S., 2004. Effects of age and alcohol exposure during early life on pyramidal cell numbers in the CA1-CA3 region of the rat hippocampus. Hippocampus 14, 124–134.

Mira, R.G., Tapia-Rojas, C., Pérez, M.J., Jara, C., Vergara, E.H., Quintanilla, R.A., Cerpa, W., 2019. Alcohol impairs hippocampal function: from NMDA receptor synaptic transmission to mitochondrial function. Drug Alcohol Depend. 205, 107628.

Mitoma, H., Manto, M., Shaikh, A.G., 2021. Mechanisms of ethanol-induced cerebellar Ataxia: underpinnings of neuronal death in the cerebellum. Int. J. Environ. Res. Public Health, 18.

Mohseni, F., Garmabi, B., Khaksari, M., 2021. Apelin-13 attenuates spatial memory impairment by anti-oxidative, anti-apoptosis, and anti-inflammatory mechanism against ethanol neurotoxicity in the neonatal rat hippocampus. Neuropeptides 87, 102130.

Molina, J.C., Chotro, G., Spear, N.E., 1989. Early (preweanling) recognition of alcohol's orosensory cues resulting from acute ethanol intoxication. Behav. Neural Biol. 51 (3), 307–325.

Montesinos, J., Pascual, M., Rodríguez-Arias, M., Miñarro, J., Guerri, C., 2016. Involvement of TLR4 in the long-term epigenetic changes, rewarding and anxiety effects induced by intermittent ethanol treatment in adolescence. Brain Behav. Immun. 53, 159–171.
Morais-Silva, G., Alves, G.C., Marin, M.T., 2016. N-acetylcysteine treatment blocks the development of ethanol-induced behavioural sensitization and related ΔFosB alterations. Neuropharmacology 110, 135–142.
Moro, L., 2020. Mitochondria at the crossroads of physiology and pathology. J. Clin. Med. 9 (6), 1971.
Murata, D., Arai, K., Iijima, M., Sesaki, H., 2020. Mitochondrial division, fusion and degradation. J. Biochem. 167 (3), 233–241.
Murphy, M.P., 2009. How mitochondria produce reactive oxygen species. Biochem. J. 417 (1), 1–13.
Nassir, F., Ibdah, J.A., 2014. Role of mitochondria in alcoholic liver disease. World J Gastroenterol: WJG 20 (9), 2136.
NIAAA-NIH (2020) https://www.niaaa.nih.gov/publications/brochures-and-fact-sheets/understanding-alcohol-use-disorder (consulted on March 2022).
Niederhofer, H., Staffen, W., Mair, A., 2003. Comparison of cyanamide and placebo in the treatment of alcohol dependence of adolescents. Alcohol Alcohol. 38 (1), 50–53.
Oliveira, S.A., Chuffa, L.G.A., Fioruci-Fontanelli, B.A., Neto, F.S.L., Novais, P.C., Tirapelli, L.F., Oishi, J.C., Takase, L.F., Stefanini, M.A., Martinez, M., Martinez, F.E., 2014. Apoptosis of Purkinje and granular cells of the cerebellum following chronic ethanol intake. Cerebellum 13 (6), 728–738.
Oliveira, A.C.A., Pereira, M.C.S., Santana, L.N.D.S., Fernandes, R.M., Teixeira, F.B., Oliveira, G.B., Fernandes, L.M.P., Fontes-Júnior, E.A., Prediger, R.D., Crespo-López, M.E., Gomes-Leal, W., Lima, R.R., Maia, C.D.S.F., 2015. Chronic ethanol exposure during adolescence through early adulthood in female rats induces emotional and memory deficits associated with morphological and molecular alterations in hippocampus. J. Psychopharmacol. 29, 712–724.
Panday, A., Sahoo, M.K., Osorio, D., Batra, S., 2015. NADPH oxidases: an overview from structure to innate immunity-associated pathologies. Cell. Mol. Immunol. 12 (1), 5–23.
Pascual, M., Baliño, P., Aragón, C.M., Guerri, C., 2015. Cytokines and chemokines as biomarkers of ethanol-induced neuroinflammation and anxiety-related behavior: role of TLR4 and TLR2. Neuropharmacology 89, 352–359.
Pautassi, R.M., Suárez, A.B., Hoffmann, L.B., Rueda, A.V., Rae, M., Marianno, P., Camarini, R., 2017. Effects of environmental enrichment upon ethanol-induced conditioned place preference and pre-frontal BDNF levels in adolescent and adult mice. Sci. Rep. 7 (1), 1–12.
Peana, A.T., Muggironi, G., Bennardini, F., 2014. Change of cystine/glutamate antiporter expression in ethanol-dependent rats. Front. Neurosci. 8, 311.
Peana, A.T., Bassareo, V., Acquas, E., 2019. Not just from ethanol. Tetrahydroisoquinolinic (TIQ) derivatives: from neurotoxicity to neuroprotection. Neurotox. Res. 36 (4), 653–668.
Pereira, R.B., Andrade, P.B., Valentão, P., 2015. A comprehensive view of the neurotoxicity mechanisms of cocaine and ethanol. Neurotox. Res. 28 (3), 253–267.
Pierce, D.R., Williams, D.K., Light, K.E., 1999. Purkinje cell vulnerability to developmental ethanol exposure in the rat cerebellum. Alcohol. Clin. Exp. Res. 23 (10), 1650–1659.
Pimentel, E., Sivalingam, K., Doke, M., Samikkannu, T., 2020. Effects of drugs of abuse on the blood-brain barrier: a brief overview. Front. Neurosci. 14, 513.
Plane, J.M., Shen, Y., Pleasure, D.E., Deng, W., 2010. Prospects for minocycline neuroprotection. Arch. Neurol. 67 (12), 1442–1448.

Qin, L., Crews, F.T., 2012. NADPH oxidase and reactive oxygen species contribute to alcohol-induced microglial activation and neurodegeneration. J. Neuroinflammation 9 (1), 1–19.
Qin, L., He, J., Hanes, R.N., Pluzarev, O., Hong, J.S., Crews, F.T., 2008a. Increased systemic and brain cytokine production and neuroinflammation by endotoxin following ethanol treatment. Neuroinflammation 5, 10.
Qin, L., He, J., Hanes, R.N., Pluzarev, O., Hong, J.S., Crews, F.T., 2008b. Increased systemic and brain cytokine production and neuroinflammation by endotoxin following ethanol treatment. J. Neuroinflammation 5 (1), 1–17.
Quertemont, E., Didone, V., 2006. Role of acetaldehyde in mediating the pharmacological and behavioral effects of alcohol. Alcohol Res. Health 29 (4), 258.
Quintanilla, M.E., Rivera-Meza, M., Berríos-Cárcamo, P., Salinas-Luypaert, C., Herrera-Marschitz, M., Israel, Y., 2016. Beyond the "first hit": marked inhibition by N-acetyl cysteine of chronic ethanol intake but not of early ethanol intake. Parallel effects on ethanol-induced saccharin motivation. Alcohol. Clin. Exp. Res. 40 (5), 1044–1051.
Quintanilla, M.E., Morales, P., Ezquer, F., Ezquer, M., Herrera-Marschitz, M., Israel, Y., 2018. Commonality of ethanol and nicotine reinforcement and relapse in Wistar-derived UChB rats: inhibition by N-acetylcysteine. Alcohol. Clin. Exp. Res. 42 (10), 1988–1999.
Quintanilla, M.E., Ezquer, F., Morales, P., Ezquer, M., Olivares, B., Santapau, D., Herrera-Marschitz, M., Israel, Y., 2020. N-acetylcysteine and acetylsalicylic acid inhibit alcohol consumption by different mechanisms: combined protection. Front. Behav. Neurosci. 14, 122.
Ramezani, A., Goudarzi, I., Lashkarboluki, T., Ghorbanian, M.T., Abrari, K., Salmani, M.E., 2012. Role of oxidative stress in ethanol-induced neurotoxicity in the developing cerebellum. Iran. J. Basic Med. Sci. 15 (4), 965.
Rao, P.S.S., Sari, Y., 2014. Effects of ceftriaxone on chronic ethanol consumption: a potential role for xCT and GLT1 modulation of glutamate levels in male P rats. J. Mol. Neurosci. 54 (1), 71–77.
Rao, P.S.S., Midde, M., N., D Miller, D., Chauhan, S., Kumar, A. and Kumar, S., 2015. Diallyl sulfide: potential use in novel therapeutic interventions in alcohol, drugs, and disease mediated cellular toxicity by targeting cytochrome P450 2E1. Curr. Drug Metab. 16 (6), 486–503.
Rivera-Meza, M., Vásquez, D., Quintanilla, M.E., Lagos, D., Rojas, B., Herrera-Marschitz, M., Israel, Y., 2019. Activation of mitochondrial aldehyde dehydrogenase (ALDH2) by ALDA-1 reduces both the acquisition and maintenance of ethanol intake in rats: A dual mechanism? Neuropharmacology 146, 175–183.
Sawant, O.B., Lunde, E.R., Washburn, S.E., Chen, W.J.A., Goodlett, C.R., Cudd, T.A., 2013. Different patterns of regional Purkinje cell loss in the cerebellar vermis as a function of the timing of prenatal ethanol exposure in an ovine model. Neurotoxicol. Teratol. 35, 7–13.
Schneider Jr., R., Santos, C.F., Clarimundo, V., Dalmaz, C., Elisabetsky, E., Gomez, R., 2015. N-acetylcysteine prevents behavioral and biochemical changes induced by alcohol cessation in rats. Alcohol 49 (3), 259–263.
Schneider, R., Bandiera, S., Souza, D.G., Bellaver, B., Caletti, G., Quincozes-Santos, A., Elisabetsky, E., Gomez, R., 2017. N-acetylcysteine prevents alcohol related neuroinflammation in rats. Neurochem. Res. 42 (8), 2135–2141.
Segura-Aguilar, J., 2019. On the role of aminochrome in mitochondrial dysfunction and endoplasmic reticulum stress in Parkinson's disease. Front. Neurosci. 13, 271.
Shih, A.Y., Johnson, D.A., Wong, G., Kraft, A.D., Jiang, L., Erb, H., Johnson, J.A., Murphy, T.H., 2003. Coordinate regulation of glutathione biosynthesis and release by Nrf2-expressing glia potently protects neurons from oxidative stress. J. Neurosci. 23 (8), 3394–3406.

Sinclair, J.D., Lindros, K.O., Terho, K., 1980. Aldehyde dehydrogenase inhibitors and voluntary ethanol drinking by rats. In: Alcohol and Aldehyde Metabolizing Systems-IV. Springer, Boston, MA, pp. 481–487.
Song, B.J., Abdelmegeed, M.A., Yoo, S.H., Kim, B.J., Jo, S.A., Jo, I., Moon, K.H., 2011. Post-translational modifications of mitochondrial aldehyde dehydrogenase and biomedical implications. J. Proteome 74 (12), 2691–2702.
Song, B.J., Akbar, M., Abdelmegeed, M.A., Byun, K., Lee, B., Yoon, S.K., Hardwick, J.P., 2014. Mitochondrial dysfunction and tissue injury by alcohol, high fat, nonalcoholic substances and pathological conditions through post-translational protein modifications. Redox Biol. 3, 109–123.
Spivak, K., Aragon, C.M., Amit, Z., 1987. Alterations in brain aldehyde dehydrogenase activity modify ethanol-induced conditioned taste aversion. Alcohol Clin. Exp. Res. 11 (6), 513–517.
Stavro, K., Pelletier, J., Potvin, S., 2013. Widespread and sustained cognitive deficits in alcoholism: a meta-analysis. Addict. Biol. 18 (2), 203–213.
Stefanatos, R., Sanz, A., 2011. Mitochondrial complex I: a central regulator of the aging process. Cell Cycle 10 (10), 1528–1532.
Stennett, B.A., Frankowski, J.C., Peris, J., Knackstedt, L.A., 2017. Ceftriaxone reduces alcohol intake in outbred rats while upregulating xCT in the nucleus accumbens core. Pharmacol. Biochem. Behav. 159, 18–23.
Strosznajder, A.K., Wójtowicz, S., Jeżyna, M.J., Sun, G.Y., Strosznajder, J.B., 2021. Recent insights on the role of PPAR-β/δ in neuroinflammation and neurodegeneration, and its potential target for therapy. NeuroMolecular Med. 23 (1), 86–98.
Surendran, S., Rajasankar, S., 2010. Parkinson's disease: oxidative stress and therapeutic approaches. Neurol. Sci. 31 (5), 531–540.
Tajuddin, N., Moon, K.H., Marshall, S.A., Nixon, K., Neafsey, E.J., Kim, H.Y., Collins, M.A., 2014. Neuroinflammation and neurodegeneration in adult rat brain from binge ethanol exposure: abrogation by docosahexaenoic acid. PLoS One 9 (7), e101223.
Takeuchi, H., 2010. Neurotoxicity by microglia: mechanisms and potential therapeutic strategy. Clin. Exp. Neuroimmunol. 1 (1), 12–21.
Tapia-Rojas, C., Mira, R.G., Torres, A.K., Jara, C., Pérez, M.J., Vergara, E.H., Cerpa, W., Quintanilla, R.A., 2017a. Alcohol consumption during adolescence: a link between mitochondrial damage and ethanol brain intoxication. Birth Defects Res. 109 (20), 1623–1639.
Tapia-Rojas, C., Pérez, M.J., Jara, C., Vergara, E.H., Quintanilla, R.A., 2017b. Ethanol consumption affects neuronal function: role of the mitochondria. Mitochondrial Dis. 14, 361–384.
Tapia-Rojas, C., Carvajal, F.J., Mira, R.G., Arce, C., Lerma-Cabrera, J.M., Orellana, J.A., Cerpa, W., Quintanilla, R.A., 2018. Adolescent binge alcohol exposure affects the brain function through mitochondrial impairment. Mol. Neurobiol. 55 (5), 4473–4491.
Tapia-Rojas, C., Torres, A.K., Quintanilla, R.A., 2019. Adolescence binge alcohol consumption induces hippocampal mitochondrial impairment that persists during the adulthood. Neuroscience 406, 356–368.
Tatemoto, K., Hosoya, M., Habata, Y., Fujii, R., Kakegawa, T., Zou, M.X., Kawamata, Y., Fukusumi, S., Hinuma, S., Kitada, C., Kurokawa, T., 1998. Isolation and characterization of a novel endogenous peptide ligand for the human APJ receptor. Biochem. Biophys. Res. Commun. 251 (2), 471–476.
Teixeira, F.B., Santana, L.N.D.S., Bezerra, F.R., De Carvalho, S., Fontes-Junior, E.A., Prediger, R.D., Crespo-Lopez, M.E., Maia, C.S.F., Lima, R.R., 2014. Chronic ethanol exposure during adolescence in rats induces motor impairments and cerebral cortex damage associated with oxidative stress. PloS one 9 (6), e101074.

Tizabi, Y., Getachew, B., Collins, M.A., 2021. Ethanol neurotoxicity. Handbook of Neurotoxicity, 1–23.

Tomlinson, D., Wilce, P., Bedi, K.S., 1998. Spatial learning ability of rats following differing levels of exposure to alcohol during early postnatal life. Physiol. Behav. 63 (2), 205–211.

Tuladhar, A.M., de Leeuw, F.E., 2019. Thalamus: a key player in alcohol use disorder and Korsakoff's syndrome. Brain 142 (5), 1170.

Uban, K.A., Sliwowska, J.H., Lieblich, S., Ellis, L.A., Wayne, K.Y., Weinberg, J., Galea, L.A., 2010. Prenatal alcohol exposure reduces the proportion of newly produced neurons and glia in the dentate gyrus of the hippocampus in female rats. Horm. Behav. 58 (5), 835–843.

Upadhyay, M., Agarwal, S., 2020. Ironing the mitochondria: relevance to its dynamics. Mitochondrion 50, 82–87.

Vasiliou, V., Pappa, A., Estey, T., 2004. Role of human aldehyde dehydrogenases in endobiotic and xenobiotic metabolism. Drug Metab. Rev. 36 (2), 279–299.

Vetreno, R.P., Crews, F., 2012. Adolescent binge drinking increases expression of the danger signal receptor agonist HMGB1 and toll-like receptors in the adult prefrontal cortex. Neuroscience 226, 475–488.

Vetreno, R.P., Qin, L., Crews, F.T., 2013. Increased receptor for advanced glycation end product expression in the human alcoholic prefrontal cortex is linked to adolescent drinking. Neurobiol. Dis. 59, 52–62.

Virgolini, M.B., Mattalloni, M.S., Deza-Ponzio, R., Albrecht, P.A., Cancela, L.M., 2019. Lead exposure and ethanol intake: oxidative stress as a converging mechanism of action. In: Neuroscience of Alcohol. Academic Press, pp. 515–525.

Vore, A.S., Deak, T., 2022. Alcohol, inflammation, and blood-brain barrier function in health and disease across development. In: International Review of Neurobiology. vol. 161. Academic Press, pp. 209–249.

Wang, Y., He, W., 2018. Endogenous mitochondrial aldehyde dehydrogenase-2 as an antioxidant in liver. In: The Liver. Academic Press, pp. 247–259.

Wilson, D.F., Matschinsky, F.M., 2020. Ethanol metabolism: the good, the bad, and the ugly. Med. Hypotheses 140, 109638.

Wu, D., Cederbaum, A.I., 2005. Oxidative stress mediated toxicity exerted by ethanol-inducible CYP2E1. Toxicol. Appl. Pharmacol. 207 (2), 70–76.

Xie, G., Hipolito, L., Zuo, W., Polache, A., Granero, L., Krnjević, K., Ye, J.H., 2012. Salsolinol stimulates dopamine neurons in slices of posterior ventral tegmental area indirectly by activating μ-opioid receptors. J. Pharmacol. Exp. Ther. 341 (1), 43–50.

Zhao, Y., Wang, C., 2015. Glu 504Lys single nucleotide polymorphism of aldehyde dehydrogenase 2 gene and the risk of human diseases. Biomed. Res. Int. 2015, 174050.

Zhou, Y., Wang, H.D., Zhou, X.M., Fang, J., Zhu, L., Ding, K., 2018. N-acetylcysteine amide provides neuroprotection via Nrf2-ARE pathway in a mouse model of traumatic brain injury. Drug Des. Devel. Ther. 12, 4117.

Zimatkin, S.M., Pronko, S.P., Vasiliou, V., Gonzalez, F.J., Deitrich, R.A., 2006. Enzymatic mechanisms of ethanol oxidation in the brain. Alcohol. Clin. Exp. Res. 30 (9), 1500–1505.

CHAPTER THREE

Neurotoxicology of nicotine and tobacco

Edward D. Levin*
Professor of Psychiatry and Behavioral Sciences, Duke University Medical Center, Durham, NC, United States
*Corresponding author: e-mail address: edlevin@duke.edu

Contents

1. Introduction

Nicotine is the primary psychoactive ingredient in tobacco. Nicotine-driven tobacco addiction is the cause of substantial toxicity, including neurotoxicity, much of which is due to the wide variety of other toxic chemicals in tobacco. However, nicotine itself can also cause neurotoxic effects. The primary determinant of nicotine neurotoxicity is developmental phase during which exposure occurs. Nicotine has long been well-characterized as a developmental neurotoxicant because cholinergic systems play critical roles controlling neurodevelopment (Abbott and Winzer-Serhan, 2012; Slikker et al., 2005; Slotkin, 2004). A great number of experimental studies have demonstrated the persisting neural and behavioral toxicity of nicotine in animal models primarily in mice and rats, but also in other species including zebrafish and monkeys. The vulnerability to nicotine neurotoxicity extends to adolescence. In adulthood there is less evidence for neurotoxicity of nicotine apart from nicotine-driven tobacco addiction and more recently addiction to e-cigarettes. There is less evidence

Advances in Neurotoxicology, Volume 8
ISSN 2468-7480
https://doi.org/10.1016/bs.ant.2022.06.003

of addiction to other forms of nicotine delivery such as gum or skin patches. In fact, there is evidence for potential therapeutic efficacy for nicotine in the treatment of cognitive impairments seen with attention deficit hyperactivity disorder (ADHD), Down's syndrome, schizophrenia and aging related cognitive impairments including Alzheimer's disease. These studies have primarily used nicotine skin patches as the delivery device. Because nicotine neurotoxicity varies so substantially over the life span, this discussion is organized by life stage. Particular neurotoxic risks are caused by nicotine during prenatal and neonatal development as well as during adolescence compared with adult and aging life phases.

2. Prenatal and neonatal development

A substantial number of studies have shown that nicotine exposure during early development causes long-term neural and behavioral toxicity (Levin and Abreu-Villaça, 2018). Prenatal nicotine in mice caused significant locomotor hyperactivity (Ajarem and Ahmad, 1998). Prenatal nicotine exposure disrupts connectivity between the thalamus and neocortex (Heath et al., 2010; King et al., 2003). Prenatal nicotine exposure in mice significantly altered the glutamate/GABA balance in the lateral dorsal tegmental nucleus which is the source of cholinergic projections controlling activity of ventral tegmental dopamine neurons (Polli and Kohlmeier, 2019). The concentration of nicotinic receptors in the brains of mice peaks during the neonatal period with a preponderance of high affinity sites and declines over the rest of the lifespan with a diversification of nicotinic binding to include both low and high affinity sites (Zhang et al., 1990).

In rats, our studies have shown that chronic nicotine exposure during gestation causes neural and behavioral dysfunction that persists into adulthood even with low levels of exposure modeling exposure to second-hand tobacco smoke. Fig. 1 shows that nicotine administered together with the complex mixture of tobacco smoke extract (TSE) produces significant locomotor hyperactivity in the offspring compared to vehicle controls as well as compered to the same dose of nicotine given alone (Hall et al., 2016). Fig. 2 shows the persisting working and reference memory impairment in the 16-arm radial maze of female offspring of dams exposed to TSE during the second half of gestation (Cauley et al., 2018). Prenatal TSE exposure also caused persistently altered cholinergic and serotonergic transmitter systems in the offspring (Slotkin et al., 2017).

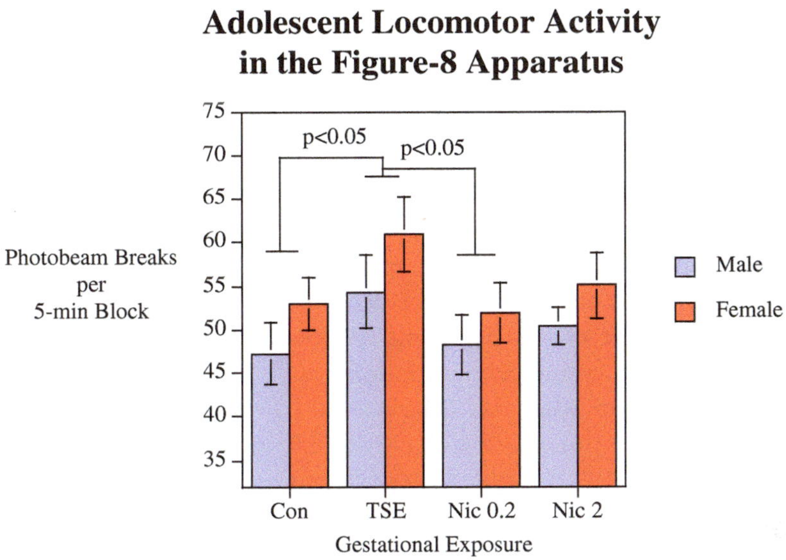

Fig. 1 Comparison of the effects of maternal exposure to nicotine alone vs nicotine in tobacco smoke extract (TSE) on offspring locomotor activity in the figure-8 maze (mean ± SEM). Low dose (0.2 mg/kg/day) nicotine in TSE caused significant locomotor hyperactivity relative to controls as well as the same dose of nicotine alone (Hall et al., 2016).

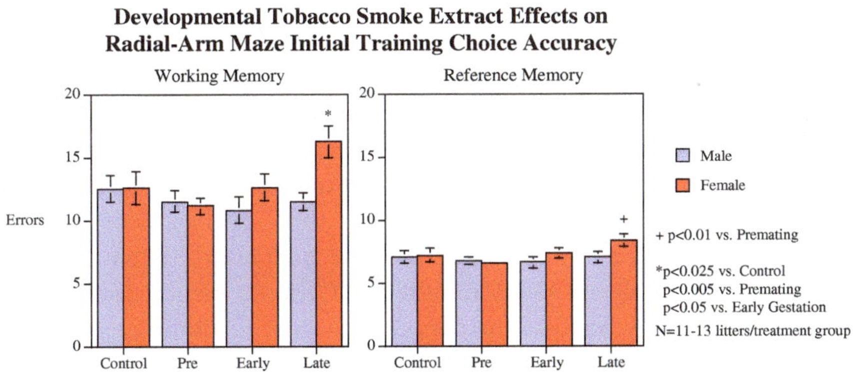

Fig. 2 Tests of effects of maternal pre-mating vs early or late gestation exposure to tobacco smoke extract (TSE) on offspring choice accuracy in the 16-arm radial maze showed that late gestational TSE exposure significantly increased both working and reference memory errors in female offspring (mean ± SEM) (Cauley et al., 2018).

The developmental neurotoxicity of nicotine generalizes beyond mammals. Complementary species can be useful in the study of developmental nicotine neurotoxicity. In zebrafish our studies have found that exposure to nicotine during the first 5 days after fertilization significantly increased startle response in adulthood (Crosby et al., 2015). Sea urchin and chick embryos are also sensitive to the developmental toxicity of nicotine (Buznikov et al., 2003; Yanai et al., 2010).

3. Adolescence

Adolescence is also a period for enhanced neurotoxicity with nicotine exposure (Korpi et al., 2015; Kwan et al., 2020; Ren et al., 2022; Slotkin, 2004). Slotkin and co-workers have demonstrated that nicotine exposure during adolescence causes persisting neurotoxic effects in rats (Slotkin, 2004). Chronic nicotine administration causes an increase in nicotinic receptor number in adult rats (Marks et al., 1983; Schwartz and Kellar, 1983). This effect is prolonged with chronic nicotine exposure in adolescent rats (Trauth et al., 1999). Adolescent male and female rats self-administered greater amounts of nicotine than rats that started in young adulthood. During adolescence males showed a greater increase than females, but females showed a more persisting increase in nicotine self-administration than males as the rats continued self-administration into adulthood (Levin et al., 2003, 2007, 2011). This may be related to the increase in impulsiveness with nicotine in adolescence (DeBry and Tiffany, 2008).

4. Adulthood

Compared to younger ages there is much less evidence of nicotine neurotoxicity in adulthood. Nicotine can at higher doses cause convulsions (Broide et al., 2002; Damaj et al., 1999; Yamamoto et al., 1966). The main neurotoxic concern of nicotine in adults is the fact that nicotine is the principal driving force underlying tobacco addiction by which other neurotoxic chemicals are self-administered. Chronic nicotine administration does temporally increase the concentrations of nicotinic acetylcholine receptors in adults (Marks et al., 1983; Schwartz and Kellar, 1983).

There have been studies of the use of nicotine skin patches to alleviate the cognitive impairments seen in adults with ADHD (Conners et al., 1996; Levin et al., 1996a), Down's syndrome (Seidl et al., 2000) and schizophrenia (Levin et al., 1996b).

5. Aging

Nicotine has been investigated as a possible therapeutic agent in aging-related cognitive impairments such as, age-associated memory impairment, mild cognitive impairment and Alzheimer's disease (Newhouse et al., 2012; White and Levin, 1999, 2004) as well as the motor dysfunction of Parkinson's disease (Tizabi et al., 2021) and symptoms of depression (McClernon et al., 2006). There is evidence for the efficacy of nicotine for improving cognition in aging (Newhouse et al., 2012; White and Levin, 1999, 2004), though further research is needed. Few adverse effects of acute or chronic nicotine skin patch treatments have been seen in chronic experimental human studies (Newhouse et al., 2012; White and Levin, 1999, 2004).

6. Future directions

The neurotoxic risks of nicotine and the complex mixture of tobacco including nicotine have been well characterized with differential effects seen over the lifespan. Of course, further research is needed for better understanding of nicotine's neurotoxicity. Paternal nicotine exposure prior to conception (2 mg/kg/day for 56 days via sc osmotic minipump) has been shown to alter sperm DNA methylation (Schrott et al., 2020) and cause long-term neurobehavioral alterations including locomotor activity and impaired habituation (Hawkey et al., 2019). Specifically, paternal nicotine exposure caused locomotor hyperactivity in male offspring, reversing the typical sex difference in locomotor activity in the figure-8 maze (Fig. 3).

Exposure to nicotine is never alone with tobacco use. Tobacco contains thousands of other compounds, some like benzo[a]pyrene (BaP) that are known neurotoxicants. We have found that prenatal exposure to nicotine + BaP causes greater persisting neural and behavioral impairments than either alone. In addition to BaP there are other polycyclic aromatic hydrocarbons that remain understudied. In addition, neurotoxic heavy metals such as cadmium are also found in tobacco (Satarug and Moore, 2004). Many studies have documented the neurotoxic effects of cadmium (Wang and Du, 2013; Wright and Baccarelli, 2007).

E-cigarettes have a much simpler chemistry, but they also can present neurotoxic risk (Ruszkiewicz et al., 2020). They contain nicotine and other potentially neurotoxic chemicals that could exacerbate the neurotoxicity of nicotine (Clapp and Jaspers, 2017; DeVito and Krishnan-Sarin, 2018). Lead has been found in the vapor output of some e-cigarettes (Eshraghian and Al-Delaimy, 2021). Of course, the neurotoxic effects of lead are very well characterized, especially with exposure during development (Grandjean and Landrigan, 2014). The interactive neurotoxic potential of nicotine + other chemicals in e-cigarettes needs further study.

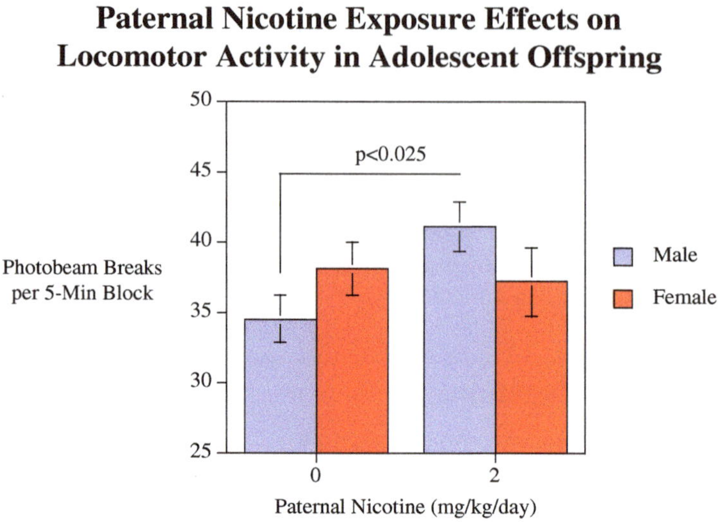

Fig. 3 Paternal pre-mating nicotine significantly increases locomotor activity in their male offspring (mean ± SEM) (Hawkey et al., 2019).

Pesticides have been found in tobacco (López Dávila et al., 2020). Pesticide neurotoxicity, especially with exposure during development, is very well characterized (Abreu-Villaça and Levin, 2017, 2018). Pesticide plus nicotine interactions have not been well enough studied. The rest of the very complex mixture of tobacco remains to be studied regarding its potentiation of nicotine's neurotoxicity. In initial studies our studies have found that nicotine together in the complex mixture of tobacco smoke extract produced greater neural and behavioral impairment than the same dose of nicotine alone (Hall et al., 2016). However, this tobacco smoke extract did not include the gaseous chemicals present in tobacco smoke. Carbon monoxide (CO) is of particular concern, given its known neurotoxic effects. There has been some earlier study of CO which also causes developmental neurotoxicity (Mactutus, 1989). However, the combined neurotoxicity of nicotine and CO and other constituents of tobacco smoke should be further studied.

In addition to the complex chemical exposure of nicotine from tobacco and e-cigarettes, nicotine is often taken together with other known neurotoxic drugs of abuse such as alcohol, opiates, stimulants and cannabis which is itself a complex mixture. How nicotine neurotoxicity might be exacerbated by these co-exposures needs further study.

Nicotine exposure might change neurotoxicity of other exposures later in life. Slotkin and colleagues found that prenatal nicotine altered neurotoxic changes to noradrenergic, serotonergic and cholinergic systems caused by later exposure to the organophosphate insecticide chlorpyrifos (Slotkin et al., 2015a,b; Slotkin and Seidler, 2015).

Nicotine is not the only nicotinic chemical that causes neurotoxicity. Neonicotinoid insecticides are becoming more widely used. Despite some research that did not find significant neurotoxicity from neonicotinoid insecticide exposure (Sheets et al., 2016) there is experimental evidence that neonicotinoid pesticide exposure like nicotine can cause long-lasting neurobehavioral toxicity after developmental exposure in zebrafish (Crosby et al., 2015) as well as neurotoxic effects in primary neural cultures from developing mice (Kimura-Kuroda et al., 2012) and lasting cognitive impairment after neonatal exposure in rats (Kara et al., 2015). Potential neurotoxic risks of neonicotinoid insecticides for potential neurotoxicity especially with early developmental exposure should be further investigated.

7. Summary and conclusions

Nicotine has proven neurotoxicity with exposure during development, both in pre and postnatal periods as well as during adolescence. The neurobehavioral toxicity caused by prenatal developmental nicotine/tobacco causes locomotor hyperactivity as well as cognitive impairment. Nicotine neurotoxicity during adulthood and aging periods of life appears to be less and there are potential therapeutic uses as a cognitive enhancer that are being investigated.

Primarily, nicotine has been characterized as a developmental neurotoxin. Nicotine exposure during prenatal development causes persisting neural and behavioral dysfunction. The rodent models shed light on the causative impact of nicotine and mechanisms of behavioral impairment. Adolescence is also a vulnerable period for nicotine neurotoxicity. Nicotinic receptor increases with chronic exposure are prolonged in adolescents and they show increased nicotine self-administration vs adults.

Acknowledgment

This research was supported by grants from the National Institutes of Health (ES022831), the U.S. Environmental Protection Agency (83543701) and the March of Dimes.

References

Abbott, L.C., Winzer-Serhan, U.H., 2012. Smoking during pregnancy: lessons learned from epidemiological studies and experimental studies using animal models. Crit. Rev. Toxicol. 42, 279–303.

Abreu-Villaça, Y., Levin, E.D., 2017. Developmental neurotoxicity of succeeding generations of insecticides. Environ. Int. 99, 55–77.

Abreu-Villaça, Y., Levin, E.D., 2018. Developmental neurobehavioral neurotoxicity of insecticides. In: Wang, C., Paule, M., Slikker W. Jr., (Eds.), Handbook of Developmental Neurotoxicology. Academic Press, San Diego, pp. 453–466.

Ajarem, J.S., Ahmad, M., 1998. Prenatal nicotine exposure modifies behavior of mice through early development. Pharmacol. Biochem. Behav. 59 (2), 313–318.

Broide, R.S., Salas, R., Ji, D., Paylor, R., Patrick, J.W., Dani, J.A., De Biasi, M., 2002. Increased sensitivity to nicotine-induced seizures in mice expressing the L250T alpha 7 nicotinic acetylcholine receptor mutation. Mol. Pharmacol. 61 (3), 695–705.

Buznikov, G.A., Slotkin, T.A., Lauder, J.M., 2003. Sea urchin embryos and larvae as biosensors for neurotoxicants. Curr. Protoc. Toxicol. Chapter 1, Unit 1.6.

Cauley, M., Hall, B.J., Abreu-Villaça, Y., Junaid, S., White, H., Kiany, A., Slotkin, T.A., Levin, E.D., 2018. Critical developmental periods for effects of low-level tobacco smoke exposure on behavioral performance. Neurotoxicology 68, 81–87.

Clapp, P.W., Jaspers, I., 2017. Electronic cigarettes: their constituents and potential links to asthma. Curr. Allergy Asthma Rep. 17 (11), 79.

Conners, C.K., Levin, E.D., Sparrow, E., Hinton, S.C., Erhardt, D., Meck, W.H., Rose, J.E., March, J., 1996. Nicotine and attention in adult attention deficit hyperactivity disorder (ADHD). Psychopharmacol. Bull. 32 (1), 67–73.

Crosby, E.B., Bailey, J.M., Oliveri, A.N., Levin, E.D., 2015. Neurobehavioral impairments caused by developmental imidacloprid exposure in zebrafish. Neurotoxicol. Teratol. 49, 81–90.

Damaj, M.I., Glassco, W., Dukat, M., Martin, B.R., 1999. Pharmacological characterization of nicotine-induced seizures in mice. J. Pharmacol. Exp. Ther. 291, 1284–1291.

DeBry, S.C., Tiffany, S.T., 2008. Tobacco-induced neurotoxicity of adolescent cognitive development (TINACD): a proposed model for the development of impulsivity in nicotine dependence. Nicotine Tob. Res. 10 (1), 11–25.

DeVito, E.E., Krishnan-Sarin, S., 2018. E-cigarettes: impact of E-liquid components and device characteristics on nicotine exposure. Curr. Neuropharmacol. 16 (4), 438–459.

Eshraghian, E.A., Al-Delaimy, W.K., 2021. A review of constituents identified in e-cigarette liquids and aerosols. Tob. Prev. Cessat. 7, 10.

Grandjean, P., Landrigan, P.J., 2014. Neurobehavioural effects of developmental toxicity. Lancet Neurol. 13, 330–338.

Hall, B.J., Cauley, M., Burke, D.A., Kiany, A., Slotkin, T.A., Levin, E.D., 2016. Cognitive and behavioral impairments evoked by low-level exposure to tobacco smoke components: comparison with nicotine alone. Toxicol. Sci. Off. J. Soc. Toxicol. 151 (2), 236–244.

Hawkey, A.B., White, H., Pippen, E., Greengrove, E., Rezvani, A.H., Murphy, S.K., Levin, E.D., 2019. Paternal nicotine exposure in rats produces long-lasting neurobehavioral effects in the offspring. Neurotoxicol. Teratol. 74, 106808.

Heath, C.J., King, S.L., Gotti, C., Marks, M.J., Picciotto, M.R., 2010. Corticothalamic connectivity is vulnerable to nicotine exposure during early postnatal development through alpha4/beta2/alpha5 nicotinic acetylcholine receptors. Neuropsychopharmacology 35 (12), 2324–2338.

Kara, M., Yumrutas, O., Demir, C.F., Ozdemir, H.H., Bozgeyik, I., Coskun, S., Eraslan, E., Bal, R., 2015. Insecticide imidacloprid influences cognitive functions and alters learning performance and related gene expression in a rat model. Int. J. Exp. Pathol. 96 (5), 332–337.

Kimura-Kuroda, J., Komuta, Y., Kuroda, Y., Hayashi, M., Kawano, H., 2012. Nicotine-like effects of the neonicotinoid insecticides acetamiprid and imidacloprid on cerebellar neurons from neonatal rats. PLoS One 7 (2), e32432.

King, S.L., Marks, M.J., Grady, S.R., Caldarone, B.J., Koren, A.O., Mukhin, A.G., Collins, A.C., Picciotto, M.R., 2003. Conditional expression in corticothalamic efferents reveals a developmental role for nicotinic acetylcholine receptors in modulation of passive avoidance behavior. J. Neurosci. 23 (9), 3837–3843.

Korpi, E.R., den Hollander, B., Farooq, U., Vashchinkina, E., Rajkumar, R., Nutt, D.J., Hyytiä, P., Dawe, G.S., 2015. Mechanisms of action and persistent neuroplasticity by drugs of abuse. Pharmacol. Rev. 67 (4), 872–1004.

Kwan, L.Y., Eaton, D.L., Andersen, S.L., Dow-Edwards, D., Levin, E.D., Talpos, J., Vorhees, C.V., Li, A.A., 2020. This is your teen brain on drugs: in search of biological factors unique to dependence toxicity in adolescence. Neurotoxicol. Teratol. 81, 106916.

Levin, E.D., Abreu-Villaça, Y., 2018. Developmental neurotoxicity of nicotine and tobacco. In: Wang, C., Paule, M., Slikker W. Jr., (Eds.), Handbook of Developmental Neurotoxicology. Academic Press, San Diego, pp. 439–452.

Levin, E.D., Conners, C.K., Sparrow, E., Hinton, S.C., Erhardt, D., Meck, W.H., Rose, J.E., March, J., 1996a. Nicotine effects on adults with attention-deficit/hyperactivity disorder. Psychopharmacology 123 (1), 55–63.

Levin, E.D., Wilson, W., Rose, J.E., McEvoy, J., 1996b. Nicotine-haloperidol interactions and cognitive performance in schizophrenics. Neuropsychopharmacology 15 (5), 429–436.

Levin, E.D., Rezvani, A.H., Montoya, D., Rose, J.E., Swartzwelder, H.S., 2003. Adolescent-onset nicotine self-administration modeled in female rats. Psychopharmacology 169, 141–149.

Levin, E.D., Lawrence, S., Petro, A., Horton, K., Rezvani, A.H., Seidler, F.J., Slotkin, T.A., 2007. Adolescent vs. adult-onset nicotine self-administration in male rats: duration of effect and differential nicotinic receptor correlates. Neurotoxicol. Teratol. 29, 458–465.

Levin, E.D., Slade, S., Wells, C., Cauley, M., Petro, A., Vendittelli, A., Johnson, M., Williams, P., Horton, K., Rezvani, A.H., 2011. Threshold of adulthood for the onset of nicotine self-administration in male and female rats. Behav. Brain Res. 225 (2), 473–481.

López Dávila, E., Houbraken, M., De Rop, J., Wumbei, A., Du Laing, G., Romero Romero, O., Spanoghe, P., 2020. Pesticides residues in tobacco smoke: risk assessment study. Environ. Monit. Assess. 192 (9), 615.

Mactutus, C.F., 1989. Developmental neurotoxicity of nicotine, carbon monoxide, and other tobacco smoke constituents. Ann. N. Y. Acad. Sci. 562, 105–122.

Marks, M.J., Burch, J.B., Collins, A.C., 1983. Effects of chronic nicotine infusion on tolerance development and nicotine receptors. J. Pharmacol. Exp. Ther. 226, 817–825.

McClernon, F.J., Hiott, F.B., Westman, E.C., Rose, J.E., Levin, E.D., 2006. Transdermal nicotine attenuates depression symptoms in nonsmokers: a double-blind, placebo-controlled trial. Psychopharmacology 189 (1), 125–133.

Newhouse, P.A., Kellar, K., Aisen, P., White, H., Wesnes, K., Coderre, E., Pfaff, A., Wilkins, H., Howard, D., Levin, E.D., 2012. Transdermal nicotine treatment of mild cognitive impairment: a six-month double-blind pilot clinical trial. Neurology 78, 91–101.

Polli, F.S., Kohlmeier, K.A., 2019. Alterations in NMDAR-mediated signaling within the laterodorsal tegmental nucleus are associated with prenatal nicotine exposure. Neuropharmacology 158, 107744.

Ren, M., Lotfipour, S., Leslie, F., 2022. Unique effects of nicotine across the lifespan. Pharmacol. Biochem. Behav. 214, 173343.

Ruszkiewicz, J.A., Zhang, Z., Gonçalves, F.M., Tizabi, Y., Zelikoff, J.T., Aschner, M., 2020. Neurotoxicity of e-cigarettes. Food Chem. Toxicol. 138, 111245.
Satarug, S., Moore, M.R., 2004. Adverse health effects of chronic exposure to low-level cadmium in foodstuffs and cigarette smoke. Environ. Health Perspect. 112 (10), 1099–1103.
Schrott, R., Rajavel, M., Acharya, K., Huang, Z., Acharya, C., Hawkey, A., Pippen, E., Lyerly, H.K., Levin, E.D., Murphy, S.K., 2020. Sperm DNA methylation altered by THC and nicotine: vulnerability of neurodevelopmental genes with bivalent chromatin. Sci. Rep. 10 (1), 16022.
Schwartz, R.D., Kellar, K.J., 1983. Nicotinic cholinergic binding sites in the brain: in vivo regulation. Science (New York, N.Y.) 220, 214–216.
Seidl, R., Tiefenthaler, M., Hauser, E., Lubec, G., 2000. Effects of transdermal nicotine on cognitive performance in Down's syndrome. Lancet 356 (9239), 1409–1410.
Sheets, L.P., Li, A.A., Minnema, D.J., Collier, R.H., Creek, M.R., Peffer, R.C., 2016. A critical review of neonicotinoid insecticides for developmental neurotoxicity. Crit. Rev. Toxicol. 46 (2), 153–190.
Slikker W., Jr., Xu, Z.A., Levin, E.D., Slotkin, T.A., 2005. Mode of action: disruption of brain cell replication, second messenger, and neurotransmitter systems during development leading to cognitive dysfunction—developmental neurotoxicity of nicotine. Crit. Rev. Toxicol. 35 (8-9), 703–711.
Slotkin, T.A., 2004. Cholinergic systems in brain development and disruption by neurotoxicants: nicotine, environmental tobacco smoke, organophosphates. Toxicol. Appl. Pharmacol. 198 (2), 132–151.
Slotkin, T.A., Seidler, F.J., 2015. Prenatal nicotine alters the developmental neurotoxicity of postnatal chlorpyrifos directed toward cholinergic systems: better, worse, or just "different?". Brain Res. Bull. 110, 54–67.
Slotkin, T.A., Skavicus, S., Levin, E.D., Seidler, F.J., 2015a. Prenatal nicotine changes the response to postnatal chlorpyrifos: interactions targeting serotonergic synaptic function and cognition. Brain Res. Bull. 111, 84–96.
Slotkin, T.A., Skavicus, S., Seidler, F.J., 2015b. Prenatal drug exposures sensitize noradrenergic circuits to subsequent disruption by chlorpyrifos. Toxicology 338, 8–16.
Slotkin, T.A., Stadler, A., Skavicus, S., Card, J., Ruff, J., Levin, E.D., Seidler, F.J., 2017. Is there a critical period for the developmental neurotoxicity of low-level tobacco smoke exposure? Toxicol. Sci. Off. J. Soc. Toxicol. 155 (1), 75–84.
Tizabi, Y., Getachew, B., Aschner, M., 2021. Novel pharmacotherapies in Parkinson's disease. Neurotoxicity Res. 39 (4), 1381–1390.
Trauth, J.A., Seidler, F.J., McCook, E.C., Slotkin, T.A., 1999. Adolescent nicotine exposure causes persistent upregulation of nicotinic cholinergic receptors in rat brain regions. Brain Res. 851 (1–2), 9–19.
Wang, B., Du, Y., 2013. Cadmium and its neurotoxic effects. Oxid. Med. Cell Longev. 2013, 898034.
White, H.K., Levin, E.D., 1999. Four-week nicotine skin patch treatment effects on cognitive performance in Alzheimer's disease. Psychopharmacology 143 (2), 158–165.
White, H.K., Levin, E.D., 2004. Chronic transdermal nicotine patch treatment effects on cognitive performance in age-associated memory impairment. Psychopharmacology 171 (4), 465–471.
Wright, R.O., Baccarelli, A., 2007. Metals and neurotoxicology. J. Nutr. 137 (12), 2809–2813.
Yamamoto, I., Otori, K., Inoki, R., 1966. Pharmacological studies on antagonists against nicotine-induced convulsions and death. Jpn. J. Pharmacol. 16 (4), 402–415.

Yanai, J., Brick-Turin, Y., Dotan, S., Langford, R., Pinkas, A., Slotkin, T.A., 2010. A mechanism-based complementary screening approach for the amelioration and reversal of neurobehavioral teratogenicity. Neurotoxicol. Teratol. 32 (1), 109–113.

Zhang, X., Wahlstrom, G., Nordberg, A., 1990. Influence of development and aging on nicotinic receptor subtypes in rodent brain. Int. J. Dev. Neurosci. 8 (6), 715–721.

CHAPTER FOUR

The neurotoxic potential of opioids including fentanyl and fentanyl analogs

R. Daniel Mellon*,†
Division of Pharmacology/Toxicology for Neuroscience, Office of New Drugs, Center for Drug Evaluation and Research, United States Food and Drug Administration, Silver Spring, MD, United States
*Corresponding author: e-mail address: Dan.Mellon@fda.hhs.gov

Contents

1. Introduction

1.1 Fentanyl and the opioid epidemic in the United States

According to the Centers for Disease Control and Prevention (CDC) website as accessed in February of 2022, nearly 841,000 people have died in the United States since 1999 from a drug overdose. Almost 73% of these deaths are attributed to opioids and most of those are attributed to synthetic opioids likely linked to illicit manufacturing rather than prescription pharmaceuticals (https://www.cdc.gov/drugoverdose/deaths/index.html).

† This publication reflects the views of the author and should not be construed to represent FDA's views or policies.

Advances in Neurotoxicology, Volume 8
ISSN 2468-7480
https://doi.org/10.1016/bs.ant.2022.04.006

The current opioid epidemic in the United States has been described as having three waves. According to the CDC, the first wave began in the early 1990s with overdose deaths from prescription opioids further increasing since 1999 (Centers for Disease Control and Prevention, 2011). The second wave began around 2010 with overdose deaths attributed to increased availability of cheaper and more potent illicit heroin in the United States (Rudd et al., 2014). Finally, the third wave began in 2013 with deaths largely attributed to synthetic opioids (Gladden et al., 2016), particularly illicitly manufactured fentanyl and fentanyl analogs that were found in combination with heroin, incorporated into counterfeit pills, and mixed with cocaine (https://www.cdc.gov/opioids/data/analysis-resources.html). Heroin and counterfeit pills sold on the street have been laced with fentanyl or fentanyl analogs and the latter made to look identical to legitimate drug products (e.g., Norco, Xanax). As such, they are not readily identified by the individuals obtaining them, and the unexpected higher potencies increase the risk of overdose. Further, cocaine users who may not be tolerant to opioids would be uniquely vulnerable to an illicit product laced with fentanyl. The large number of fentanyl analogs with limited or no data on metabolic profiles make routine detection of these compounds either by law enforcement or medical professions challenging and likely result in an underestimation of the exposures to these compounds (Armenian et al., 2018).

Citing an imminent hazard to the public safety, in February of 2018 the United States Drug Enforcement Agency temporarily scheduled all fentanyl related substances that were not already listed in the Controlled Substances Act and their isomers, esters, ethers, salts and salts of isomers, esters, and ethers into Schedule I. This permitted all regulatory controls and civil and criminal sanctions applicable to Schedule I substances to be imposed on anyone who handles or proposes to handle fentanyl-related substances (Drug Enforcement Administration, 2018).

There are very little data on the potency, selectivity, or toxicity of the fentanyl analogs potentially present in illicit drugs. As such, the ability of the user to predict the dose needed to obtain the desired effects and avoid unintended toxicity is very limited. Wilde and colleagues reviewed the literature to obtain data on the relative potency of 23 different fentanyl analogs and to begin to characterize the metabolic profile of these compounds. These data can also help with identification of illicit opioids in both confiscated materials and in people presenting in an emergency department with a suspected overdose, to assist in their diagnosis of opioid overdose and treatment. The authors noted that the literature supports the conclusion that carfentanil is approximately 30 to 100 times more potent than fentanyl and several other compounds show potencies greater than fentanyl (e.g., furanylfentanyl and ocfentanil were reported as approximately 7 and 2.5 times more potent than fentanyl, respectively). Potency estimates were not available for all of the other compounds. Many of them were less potent than fentanyl but with varying degrees of potency (Wilde et al., 2019).

Varshneya and colleagues noted that many of the fentanyl analogs identified in illicit products have never been evaluated in vivo to determine their pharmacological effects. To begin to address this data gap, they tested seven compounds in mice that have either been reported in internet forums or identified in confiscated materials (isobutyrylfentanyl, crotonylfentanyl, valerylfentanyl, *para*-fluorobutyrylfentanyl, *para*-methoxybutyrylfentanyl, thiophenefentanyl, and benzodioxolefentanyl) to compare relative potency to morphine and fentanyl in two well-known opioid pharmacodynamic endpoints (i.e., locomotor activity and anti-nociception), and tested the ability of any effects to be antagonized by naltrexone. For several of these compounds, these were the first in vivo data published. The results demonstrated clear differences among compounds. Isobutyrylfentanyl and para-methoxybutyrylfentanyl were 1.04 times and 0.758 times, respectively, as potent as fentanyl in the antinociception assay. The other compounds showed lower potency than fentanyl and benzodioxolefentanyl was less potent than morphine. Further, the antinociceptive effects of all tested substances were naltrexone sensitive and therefore, at least in part, mediated by opioid receptors (Varshneya et al., 2019). These same investigators later reported comparable data on 7 additional fentanyl analogs (beta-methylfentanyl, *para*-methylfentanyl, *para*-methoxyfentanyl, fentanyl carbamate, 3-furanylfentanyl, phenylfentanyl, and beta'-phenylfentanyl) and note that although the effects are attenuated by naltrexone, the differences in the ability of naltrexone to antagonize the effects suggest that there may be more than opioid receptors mediating the effects of these compounds (Varshneya et al., 2021). In their latest work, Varshneya and colleagues examined the respiratory depressant effects of seven of the previously studied 14 fentanyl analogs (isobutyrylfentanyl, crotonylfentanyl, *para*-methoxyfentanyl, *para*-methoxybutyrylfentanyl, 3-furanylfentanyl, thiophenefentanyl, and benzodioxolefentanyl) in their mouse models and compared them to morphine, fentanyl, and buprenorphine. All compounds demonstrated the capacity to produce respiratory depression in this model that was attenuated by naloxone. All but benzodioxolefentanyl were more potent than morphine and none were as potent as fentanyl. These results illustrate the significant variability in potencies of these compounds (Varshneya et al., 2022).

1.2 History of therapeutic fentanyl and related substances

Despite the widespread misuse of opioids, opioids, including fentanyl and approved fentanyl analogs have a long history of safety and efficacy in humans when used appropriately and in accordance with prescription drug product labeling. Fentanyl, originally code named Compound R 4263, was first synthesized in 1960 by Dr. Paul Janssen, the founder of Janssen Pharmaceutical (Beerse, Belgium), and described in the published literature in 1962 (Janssen, 1962). Dr. Janssen's intent was to create a potent, rapid-acting analgesic. He started with the chemical structure of meperidine, a phenylpiperidine structure, rather than morphine, a phenanthrene structure, because the phenylpiperidine structure of meperidine was less complex and easier to modify. Ultimately, he was able to identify a series of drugs in this class which demonstrated high potency relative to morphine and greater lipophilicity which contributes to rapid onset (Stanley, 1992, 2014). Fentanyl was reported to be approximately 100 times more potent than morphine as an analgesic and, because it also has significantly greater lipophilicity, it demonstrated rapid onset which makes it very useful as an analgesic in the context of surgical and post-surgical pain control.

Fentanyl was first used as an intravenous analgesic in Europe as early as 1963 and the published literature includes early reports of its utility as part of an anesthetic regimen in the United States as early as 1964 (Corssen et al., 1964). Prior to fentanyl, morphine was used for anesthesia during cardiac surgery because it produced cardiovascular stability; however, there were some disadvantages, including occasional patient awareness, hypotension possibly from histamine release, and hypertension probably from catecholamine release. In contrast, fentanyl was able to produce complete anesthesia with no evidence of recall by the patient and greater cardiovascular stability during cardiac surgery (Sebel et al., 1981).

The structures of morphine, heroin, and compounds in the fentanyl series are depicted in Table 1 along with their relative potency and lipophilicity compared to morphine [data sources include the FDA-approved drug product labeling, ChemIDPlus Online, and published reviews (O'Brien, 2017; Yaksh and Wallace, 2017)]. Lipophilicity can be measured by the ability of a compound to partition into octanol versus water. Compounds with a value of 1 equally distribute between octanol and water. Compounds with a value greater than 1 distribute into octanol more than water and the higher the value the greater the lipophilicity. Compounds with high lipophilicity are expected to more rapidly enter

Table 1 Chemical structures, potency, lipophilicity, and duration of action estimates of typical opioids with a focus on fentanyl analogs.

Drug substance	Structure	Potency relative to morphine (IV)	Lipophilicity (Log P; octanol:water)	Typical duration of action (IV) in humans
Morphine	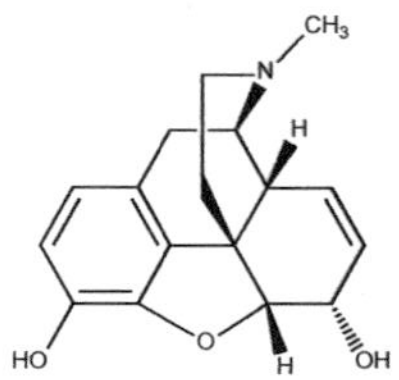	1	0.89	3–7 h
Heroin	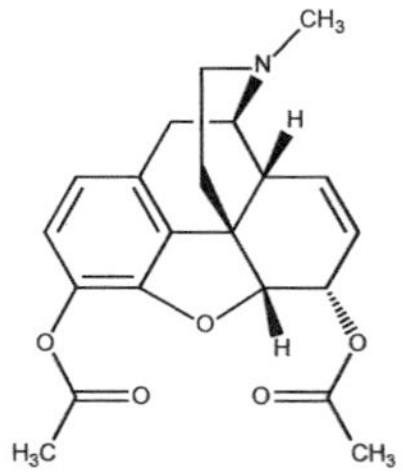	~2-4 ×	1.58	3–5 h
Meperidine	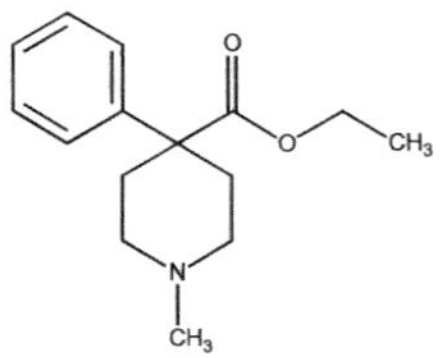	~0.13–0.1 ×	2.72	1.5–3 h
Fentanyl	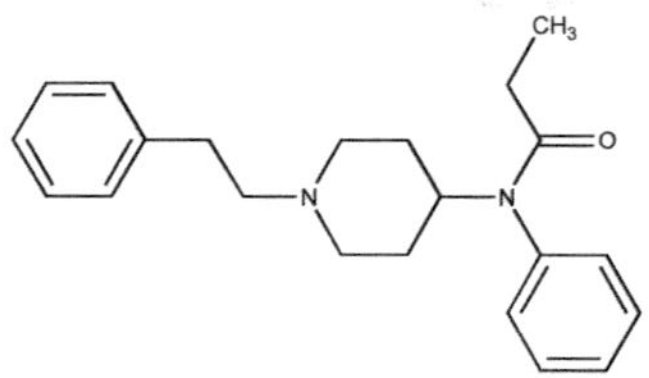	~50-100 ×	4.05	30–60 min

Continued

Table 1 Chemical structures, potency, lipophilicity, and duration of action estimates of typical opioids with a focus on fentanyl analogs.—cont'd

Drug substance	Structure	Potency relative to morphine (IV)	Lipophilicity (Log P; octanol:water)	Typical duration of action (IV) in humans
Alfentanil	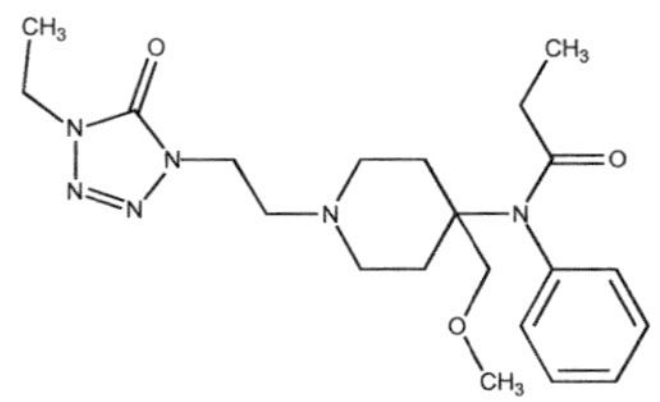	~10 ×	2.16	15 min
Sufentanil	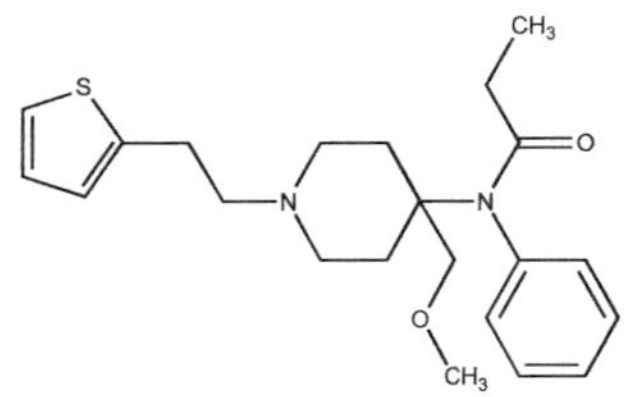	~1000 ×	3.95	20 min
Remifentanil	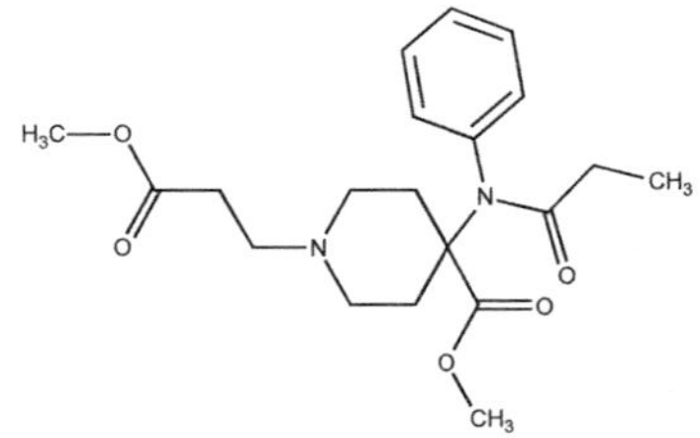	~100 ×	1.9[a]	< 10 min

Carfentanil (Veterinary)	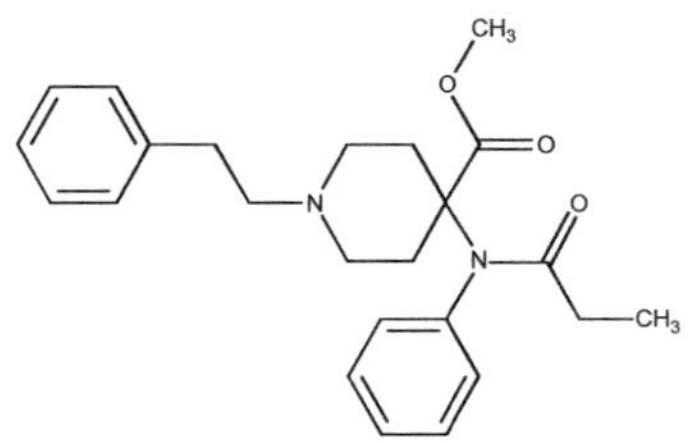	~10,000 ×	3.8[a]	NA
Thiafentanil (Veterinary)	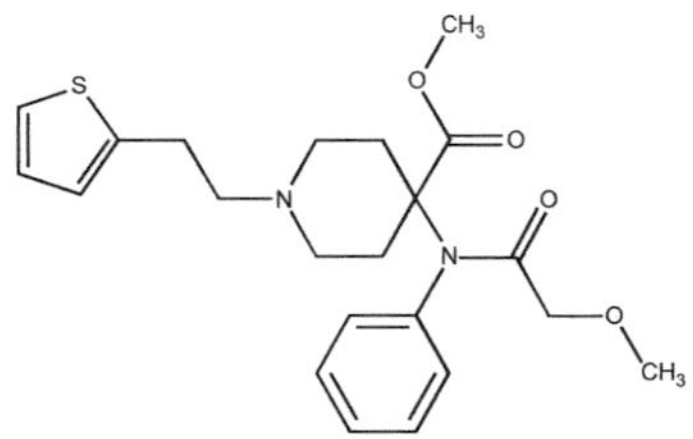	~6000 ×	3.1[a]	NA
Etorphine (Veterinary)	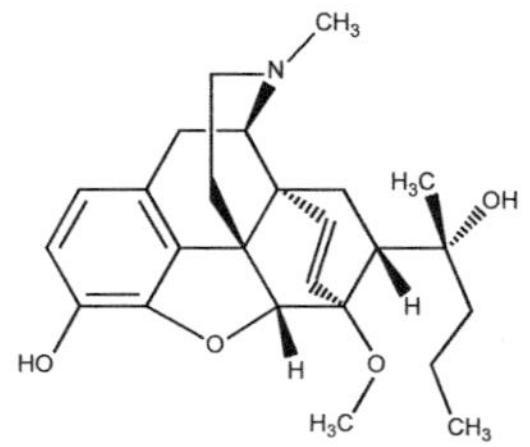	~6000 ×	2.79	NA

[a]Estimated Log P by PubChem.
NA = Definitive human data are not available.

the brain and more rapidly redistribute to fat compared to compounds with lower lipophilicity, which tend to stay in the blood stream and distribute less. In general, rapid redistribution can result in a shorter duration of action (smaller alpha half-life) although factors such as metabolism and elimination also contribute to ultimate duration of effect.

Fentanyl was first approved by the United States Food and Drug Administration on February 19, 1968 with the brand name Sublimaze®, an injectable form of the drug. Once introduced, fentanyl replaced morphine as an analgesic for use during surgical procedures because of its preferred pharmacokinetic properties (fast onset and fast offset) and reduced histamine releasing effects. As per the approved labeling, Sublimaze currently has the following indications:

- For analgesic action of short duration during the anesthetic periods, premedication, induction and maintenance, and in the immediate post-operative period (recovery room) as the need arises
- For use as a narcotic analgesic supplement in general or regional anesthesia
- For administration with a neuroleptic as an anesthetic premedication, for the induction of anesthesia and as an adjunct in the maintenance of general and regional anesthesia
- For use as an anesthetic agent with oxygen in selected high-risk patients, such as those undergoing open heart surgery or certain complicated neurological or orthopedic procedures

As evident from the indications, Sublimaze is only labeled for use in a highly-monitored clinical setting where interventions such as opioid antagonists and trained anesthetists are available to help maintain the airway.

Duragesic® (fentanyl) transdermal system was approved by the FDA in 1990 for the management of pain in opioid tolerant individuals who require daily, around-the-clock opioids and for whom alternative treatment options are inadequate. It is currently labeled with a "limitations of use" (LOU) statement that notes the risk of addiction, abuse, and misuse and that the product should not be used as an "as-needed (prn) analgesic." This was the first fentanyl drug product that was indicated for use in a less monitored setting. Actiq® (fentanyl citate) oral transmucosal lozenge was the first fentanyl drug product approved for an indication of breakthrough cancer pain. Multiple drug products were subsequently approved for a similar indication as summarized in Table 2 below. Because of the potential for diversion, these drugs are approved in the United States with a Risk Evaluation and Mitigation Strategy (REMS) to help ensure that the benefits of a drug

Table 2 Chronology of FDA-approved fentanyl series NDA drug products.

Drug substance	Brand named drug product(s)	Date of FDA approval	Dosage form (and Route of Administration)	Indications
Fentanyl	Sublimaze®	2/19/1968	Injection (IV, IM)	• For analgesic action of short duration during the anesthetic periods, premedication, induction and maintenance, and in the immediate postoperative period (recovery room) as the need arises. • For use as an opioid analgesic supplement in general or regional anesthesia. • For administration with a neuroleptic as an anesthetic premedication, for the induction of anesthesia and as an adjunct in the maintenance of general and regional anesthesia. • For use as an anesthetic agent with oxygen in selected high-risk patients, such as those undergoing open heart surgery or certain complicated neurological or orthopedic procedures.
Sufentanil	Sufenta®	5/4/1984	Injection (IV, epidural)	• As an analgesic adjunct in the maintenance of balanced general anesthesia in patients who are intubated and ventilated. • As a primary anesthetic agent for the induction and maintenance of anesthesia with 100% oxygen in patients undergoing major surgical procedures, in patients who are intubated and ventilated, such as cardiovascular surgery or neurosurgical procedures in the sitting position, to provide favorable myocardial and cerebral oxygen balance or when extended postoperative ventilation is anticipated. • For epidural administration as an analgesic combined with low dose (usually 12.5 mg per administration) bupivacaine usually during labor and vaginal delivery.

Continued

Table 2 Chronology of FDA-approved fentanyl series NDA drug products.—cont'd

Drug substance	Brand named drug product(s)	Date of FDA approval	Dosage form (and Route of Administration)	Indications
Alfentanil	Alfenta®	12/29/1986	Injection (IV)	• As an analgesic adjunct given in incremental doses in the maintenance of anesthesia with barbiturate/nitrous oxide/oxygen. • As an analgesic administered by continuous infusion with nitrous oxide/oxygen in the maintenance of general anesthesia • As a primary anesthetic agent for the induction of anesthesia in patients undergoing general surgery in which endotracheal intubation and mechanical ventilation are required. • As the analgesic component for monitored anesthesia care (MAC)
Fentanyl	Duragesic®	8/7/1990	Transdermal	• For the management of pain in opioid-tolerant patients, severe enough to require daily, around-the-clock, long-term opioid treatment and for which alternative treatment options are inadequate. (LOU)
Remifentanil	Ultiva®	7/12/1996	Injection (IV)	• As an analgesic agent for use during the induction and maintenance of general anesthesia for inpatient and outpatient procedures. • For continuation as an analgesic into the immediate postoperative period in adult patients under the direct supervision of an anesthesia practitioner in a postoperative anesthesia care unit or intensive care setting. • As an analgesic component of monitored anesthesia care in adult patients.

Fentanyl	Actiq®	11/4/1998	Transmucosal	• For the management of breakthrough pain in cancer patients 16 years of age and older who are already receiving and who are tolerant to around-the-clock opioid therapy for their underlying persistent cancer pain. (LOU)
	Fentora®	9/25/2006	Tablet (Buccal, Sublingual)	• For the management of breakthrough pain in patients with cancer who are already receiving and who are tolerant to opioid therapy for their underlying persistent cancer pain.
	Onsolis®	7/16/2009 (discontinued)	Film (Buccal)	
	Abstral®	1/7/2011	Tablet (Sublingual)	
	Lazanda®	6/30/2011	Spray, Metered (Nasal)	
	Subsys®	1/4/2012	Spray (Sublingual)	• For the management of breakthrough pain in cancer patients 18 years of age and older who are already receiving and who are tolerant to opioid therapy for their underlying persistent cancer pain. Patients must remain on around-the-clock opioids when taking SUBSYS. (LOU)
Sufentanil	Dsuvia®	11/2/2018	Tablet (Sublingual)	• For use in adults in a certified medically supervised healthcare setting, such as hospitals, surgical centers, and emergency departments, for the management of acute pain severe enough to require an opioid analgesic and for which alternative treatments are inadequate. (LOU)

outweigh its risk. These can be accessed at the FDA website (search Rems@FDA). To date, there are three main shared REMS programs for opioids that include both innovator and generic drug products: The Extended-Release Long Acting (ER/LA) REMS (aka ER/LA REMS) was expanded in 2018 to include immediate-release opioids and is now referred to as the Opioid Analgesic REMS, the Transmucosal Immediate-Release Fentanyl REMS (aka TIRF REMS), and the Buprenorphine-Containing Transmucosal products for Opioid Dependence REMS (aka BTOD REMS).

Sufentanil and alfentanil were also discovered by Dr. Janssen and his colleagues. Sufentanil (R 30730) was first reported in the literature in 1976 (Niemegeers et al., 1976), approved by FDA in 1984, and marketed as being more potent than fentanyl and with a slightly faster onset. Alfentanil (R 39209) was initially reported in the literature in 1979 and further described in 1980 (de Castro et al., 1979a; Kay and Pleuvry, 1980), approved by FDA in 1986, and marketed as having the fastest onset and shortest duration of action of the three compounds. Remifentanil (GI 87084 ×) was discovered by Dr. Paul Feldman and colleagues at Glaxo and first reported in the literature in 1991 (Feldman et al., 1991). It was approved by the FDA in 1996 and marketed as an ultra-short acting opioid analgesic that was not metabolized by the liver.

Carfentanil, thiofentanil, and etorphine are not approved for human use, are used only in veterinary practice for the sedation of large wild animals (Zuba and Greenberg, 2019), and are typically administered intramuscularly via a dart (Zawilska et al., 2021). Because of this veterinary use, DEA classified these compounds as Schedule II. Data in humans are limited; however human data have been generated for carfentanil in part because $[^{11}C]$ carfentanil has been used experimentally as a radiotracer for PET imaging (Cumming et al., 2019) and more recently because carfentanil has been detected in illicit opioids and has contributed to increased mortality of users given its extreme potency. The data on the half-life of carfentanil in humans after intravenous use is extremely limited but estimates range from 35.4 min after a dose of 1 μg/kg (Minkowski et al., 2012) to over 5 h in recreational users potentially due to nonlinear kinetics (Uddayasankar et al., 2018).

1.2.1 Opioid-induced respiratory depression

In addition to the well-known beneficial pharmacodynamic effects of opioids as extremely effective analgesics, anesthetics, and antidiarrheal mediations, the two most characterized adverse pharmacodynamic effects of opioids are respiratory depression and addiction. Detailed discussion of the central pathways involved in these effects is beyond the scope of this chapter; however, a brief discussion will set the stage for the subsequent discussion of how these known effects of opioids contribute to adverse outcomes in the setting of overdose and misuse.

There are three main subtypes of opioid receptors, designated mu, delta, and kappa. Activation of mu opioid receptors is believed to produce the primary intended clinical effects of opioid agonists. All mu opioid receptor agonists are also known to produce respiratory depression via combined depression of the complex respiratory rhythm generation network in the ventrolateral medulla of the brain, depression of the ventilatory response to increased carbon dioxide, and an effect on chemosensors in the brainstem that measure CO_2 and in the carotid and aortic body that respond to hypoxia (Ramirez et al., 2021; Yaksh and Wallace, 2017). Most opioid-related overdose deaths are the result of respiratory depression. The time to onset and duration of effect of respiratory depression following intravenous administration depends largely on the lipophilicity of the drug substance. As stated above, highly lipophilic compounds have rapid onset and generally shorter duration of effect which makes timely awareness of overdose conditions and ability to intervene with opioid antagonist treatments challenging. Less lipophilic compounds tend to have slower onset of effects but longer durations of effect potentially providing more time for diagnosis and treatment for overdose. However, longer acting opioids have greater potential for "renarcotization" particularly given the shorter half-life of naloxone overdose treatments. The effects of fentanyl and fentanyl analogs on respiration and ability to treat opioid overdose may be further complicated by their ability to produce rigidity in the diaphragm and chest wall that has been hypothesized to contribute to the increased mortality in subjects unknowingly self-administering fentanyl or fentanyl analogs instead of their usual conventional opioid (Burns et al., 2016; Pergolizzi et al., 2021). This phenomenon is well known to anesthesiologists who refer to the effect as wooden chest syndrome; however, this phenomena is not as well known by individuals misusing opioids (Davis and Behm, 2020).

Although opioids can result in mortality due to respiratory depression, toxicity from opioid overdoses can result in other adverse effects that can have a significant impact on the quality of life and overall public health. For example, brain deprivation of oxygen and glucose for even relatively short periods of time can result in significant tissue injury including neuronal necrosis. The brain is particularly vulnerable to perturbations in oxygen and glucose levels for several reasons, first is the fact that, compared to other tissues, the brain has extremely high energy demand in the form of glucose coupled with a small reserve of glucose. Second, injury to the complex network of axons and dendrites can extend retrogradely to remote neurons and lead to "disuse atrophy" of neurons as a result of depletion of neuronal input. Third, the brain is basically a hollow organ with a ventricular system in the center and the major arteries feeding the brain are primarily on the surface with arteries that penetrate the parenchyma of smaller diameter resulting in increased vulnerability to perturbations in adequate blood flow. Finally, because the brain is encased by a protective skull, blood flow to the parenchyma of the brain can also be adversely impacted by edema due to pressure build up inside the skull (Kuroiwa and Okeda, 1994).

The subcategory of hypoxia associated with opioid overdose is commonly referred to as anoxia or hypoxic hypoxia and is the result of an overall decrease in blood gas exchange in the lung. The impact of anoxia and hypoxic hypoxia may differ depending on the inciting agent, as neuronal damage can be influenced not only by direct depletion of oxygen and glucose but indirectly via alterations in cerebral blood flow as a result of the toxic agent causing the hypoxic state. Common findings of anoxic and hypoxic hypoxia include severe congestion and swelling of the brain, neuronal changes as a result of ischemia or neuronal loss, complete or focal infarcts of the cerebral white matter and/or palladium, infarcts of the cerebral cortex or basal ganglia, or ulegyria (shrunken and flattened cortex). Specifically vulnerable tissues include the pyramidal cells of the third layer of the cerebral cortex, neurons in CA1 and CA3 layers of the hippocampus, and neurons of the caudate and putamen, and Purkinje cells (Kuroiwa and Okeda, 1994).

Milroy and Parai summarized common histopathological changes induced by drugs of abuse. They note that the most common CNS finding in opioid-related deaths is hypoxia-related brain damage, most commonly noted in the hippocampus and cerebellum, areas of the brain specifically vulnerable to hypoxia (Milroy and Parai, 2011). They also note that, although rare, histopathological changes consistent with spongiform leukoencephalopathy have been reported in patients using heroin. Histopathologically, this manifests as brain edema with white matter vacuolization, and decreased myelin around cavities formed by coalescence of vacuoles. There is usually no evidence of inflammatory cells. Reports of opioid-related leukoencephalopathies are summarized and discussed later in the chapter.

2. Neurotoxicology of fentanyl and related compounds

To begin to evaluate the neurotoxic potential of fentanyl and related compounds in light of the current opioid epidemic, a comprehensive literature search was conducted focusing on the discovery and initial characterization of fentanyl with specific emphasis on reports that characterized adverse neurological effects indicative of signals for neurotoxicity. Initial publications described nonclinical studies reporting both seizures and neuronal damage. Although fentanyl and fentanyl analogs appear to be major contributors to the current wave of opioid overdoses, published reports describing data from other opioids were also included as these frequently augment the published findings with fentanyl and contribute to our understanding of potential class effects. These are reviewed largely in chronological order as they help to present the overall historical understanding of the potential neurotoxicity of fentanyl and related opioids. Our understanding of many of these observations remain incomplete and continue to evolve.

2.1 Early reports of seizures and resulting neuropathology

2.1.1 Animal data

The neurotoxic potential of fentanyl was reported in the literature as early as 1967. These early reports focused on the potential for fentanyl to induce seizures and alter blood flow in the CNS in nonclinical models. Freeman and Ingvar reported that low doses of intravenous fentanyl did not significantly alter cortical blood flow in the brain stem of intact cats, but high doses increased cortical blood flow suggesting increased metabolism. There was also an increase in seizure risk, which was mediated through the reticular brain stem based on results of brainstem transection studies. The dose of fentanyl necessary to produce evidence of seizures in the cat was reported to be 20–80 μg/kg (Freeman and Ingvar, 1967). Plasma fentanyl levels were not provided; however, based on dose normalized to body surface area (assuming a mean body weight in the cats of 3 kg) the human equivalent doses are estimated to be 7.4–29.8 μg/kg (assumes a 60 kg person).

De Castro and colleagues studied the effects of various opioids including high doses of fentanyl in the dog to help characterize the potential adverse effects of anesthetic doses of these drugs. They noted that the high doses of morphine necessary to produce deep surgical anesthesia were accompanied by severe tachycardia, vasodilation, and depression of myocardial contractility. In contrast, fentanyl and sufentanil produced a better cardiovascular profile (de Castro et al., 1979b). This group also reported that opioids, including fentanyl and sufentanil could produce seizures in this model. The threshold dose of intravenous fentanyl necessary to produce epileptiform EEG changes in this model was 4000 μg/kg (de Castro et al., 1979c). This dose corresponds to a human equivalent dose of 2222 μg/kg person based on body surface area. Finally, the research team reported that opioids, including high-dose fentanyl, increased oxygen consumption, increased catecholamines, and produced a state of metabolic acidosis and hypermetabolism (de Castro et al., 1979d).

Carlsson and colleagues reported that intravenous fentanyl (100 μg or 16 μg/kg human equivalent dose based on body surface area) administration to rats over a five-minute period followed by continuous infusion at the rate of twice the initial dose per hour decreased both cerebral blood flow and cerebral metabolic rate. Doses of 200 and 400 μg/kg resulted in seizures documented via EEG recordings under the conditions tested (32 and 64 μg/kg human equivalent dose based on body surface area). Seizure activity was reversed by naloxone (Carlsson et al., 1982).

Tommasino and colleagues also reported that high doses of intravenous fentanyl (200 and 400 μg/kg) administration to the rat produced spike wave patterns on EEG indicative of seizures. Under the conditions tested, these doses also produced a significant decrease in local cerebral glucose metabolism with the greatest effects noted in the limbic system (Tommasino et al., 1984). Plasma C_{max} or AUC values were not reported. However, these doses correspond to 32 and 64 μg/kg human equivalent dose based on body surface area which are significantly higher than the 2 μg/kg recommended low dose or the 2 to 20 μg/kg moderate doses recommended in the FDA-approved Sublimaze labeling. However, based on body surface area comparisons, the doses tested are within the range of the high doses listed in the Sublimaze label (20 to 50 μg/kg) that are used during open heart surgery or more complicated neurosurgical or orthopedic procedures and the doses listed for use as a general anesthetic (i.e., doses up to 150 μg/kg). Further studies by this group reported that local cerebral blood flow in the rat increased during high-dose fentanyl-induced EEG spike wave patterns in all 29 brain structures evaluated with statistically significant increases in the superior colliculus, sensorimotor cortex, and pineal body. Under the conditions tested, local cerebral vascular resistance was decreased. With onset of EEG recorded seizure, local cerebral blood flow increased by 58% to 231% of control values in all but 5 of the 29 brain structures examined. During post seizure EEG, local cerebral blood flow remained elevated in all structures with increases ranging from 80% of controls in the cerebellar white matter to over 350% of controls in some structures such as the septal nucleus, hippocampus, and hypothalamic areas. The mean fentanyl dose reported in this rat study to produce spike patterns was 89 μg/kg and to produce seizures was 213 μg/kg (Maekawa et al., 1984). These rat doses correspond to human equivalent doses of 14.3 and 34 μg/kg based on body surface area. These doses exceed the recommended low doses for analgesia but are within doses used for general anesthesia and, in theory, may be reached during fentanyl overdose scenarios without the supportive care and concomitant medications that occur in the clinical setting.

In the first study testing alfentanil in a rat model, Kofke and colleagues reported evidence of limbic system seizures with hypermetabolism and neurological lesions after high doses of alfentanil (150 or 1000 μg/kg). Exposure data were not reported; however, these doses correspond to human equivalent doses of 24 and 161 μg/kg based on body surface area. The induction dose reported in the FDA-approved labeling for Alfenta is between 8 and 20 μg/kg for spontaneous breathing or assisted ventilation or 20–50 μg/kg for assisted or controlled ventilation, suggesting these acute doses were higher than standard anesthetic doses without measures taken to ensure adequate oxygenation. In a separate experiment, alfentanil doses of 2000 μg/kg (322.5 μg/kg human equivalent dose based on body surface area) over 1–2 min followed by an infusion of 33.3 μg/kg/min for 1 h resulted in limbic system seizure activity and histopathological evidence of hippocampal and amygdaloid damage (Kofke et al., 1992).

Kofke and colleagues also published results of a rodent model testing convulsant doses of fentanyl, alfentanil, and sufentanil specifically examining the potential for these drugs to induce neurotoxicity (Kofke et al., 1996a). Ventilated rats were treated with either a loading dose of fentanyl 1000 μg/kg followed by a maintenance dose of 40 μg/kg/min, a loading dose of sufentanil of 400 μg/kg followed by a maintenance dose of 13.3 μg/kg/min, or a loading dose of alfentanil of 1500 μg/kg followed by a maintenance dose of 150 μg/kg/min. Loading doses were administered over 1–2 min and maintenance doses were administered for 2 h. Opioid treatments resulted in evidence of neuropathology primarily in the cortical regions and limbic system (e.g., anterior piriform cortex, entorhinal cortex, adjacent amygdala, and hippocampus). To determine if this damage was a result of hypertension, hexamethonium, a peripherally restricted ganglionic antagonist was administered during the infusion in some animals. Hexamethonium treatment was reported to attenuate the neuropathology in this model. Plasma AUC and C_{max} values were not reported; however, the doses of the opioids tested are higher than those recommended in FDA labeling based on body surface area comparison. Specifically, a loading dose of 1000 µg/kg fentanyl is equivalent to a human dose of 161 µg/kg, a dose of 400 µg/kg sufentanil is equivalent to 64.5 µg/kg human dose, and a dose of 1500 µg/kg alfentanil in the rat is equivalent to 242 µg/kg in the human based on body surface area.

To evaluate the dose dependency of fentanyl-induced neurotoxicity in their rat model, Kofke and colleagues administered loading doses ranging from 50 to 3200 µg/kg followed by maintenance doses ranging from 2 to 128 µg/kg/min for 2 h. No lesions were noted in animals treated with up to 200 µg/kg fentanyl as a loading dose with 8 µg/kg/min maintenance dose for 2 h. Evidence of neuropathology, primarily in the limbic system, was noted in animals receiving loading-maintenance doses of 400–16, 800–32, 1600–64, and 3200–18 µg/kg-mcg/kg/min (Kofke et al., 1996b). Again, pharmacokinetic exposure data were not reported; however, a loading dose of 400 µg/kg of fentanyl in the rat is equivalent to 64.5 µg/kg in the human based on body surface area, which is higher than most recommended analgesic doses of Sublimaze as per FDA-approved labeling but within the labeled dosing recommendations for some anesthetic uses. The no adverse effect level loading dose of 200 µg/kg fentanyl corresponds to a human equivalent dose of 32 µg/kg, which is within the recommended dosing recommendations as per Sublimaze labeling.

In a subsequent study, Kofke and colleagues provided data that fentanyl can exacerbate damage from forebrain ischemia in their rat model. Interestingly, in this model fentanyl-related increase in neuronal degeneration in the presence of ischemia was primarily in the caudate/putamen rather than the limbic system. The mechanism mediating this exacerbated effect is not clear but the authors speculated that it could be due to an increase in brain temperature or other neurotransmitter-based effects rather than hypoxia or seizures (Kofke et al., 1999). To explore the role of other neurotransmitters, Kofke and colleagues examined the effect of various pharmacological agents on fentanyl-induced neurotoxicity in their rat model. They report that drugs that reduced the effects of acetylcholine and catecholamines attenuated fentanyl-induced neurotoxicity without altering fentanyl-induced changes in EEG and they suggest that drugs that may alter glutamatergic and histamine receptor activation may also play a role in this response. However, no clear single mechanism emerged to explain these findings (Kofke et al., 2000).

Collectively, these reports suggest that, under the conditions tested, fentanyl and related compounds can increase energy metabolism of CNS tissues in the rodent model and produce seizures based on both clinical signs and EEG recordings. They also suggest that different animal species may show differential sensitivity to fentanyl-induced seizures, with cats showing the greatest sensitivity followed by rats, and dogs showing the lowest sensitivity, based on body surface area comparisons. These observations are also consistent with studies reporting that central endogenous opioid pathways playing a role in seizure activity in animal models (Frenk, 1983; Frenk et al., 1978a,b; Snead, 1986; Urca et al., 1977). Studies with morphine, however, suggest low doses have an anticonvulsive effect but that higher doses have a proconvulsive effect (Frenk, 1983). It has been speculated that at these higher doses morphine may also be acting at receptors other than the mu opioid receptor, such as the delta opioid receptor or other targets (Frenk et al., 1982). Further, published nonclinical studies suggest that opioids may reduce the threshold for seizures induced by pentylenetetrazole (Frenk, 1983; Przewlocka et al., 1983; Shafaroodi et al., 2004). Published in vitro studies also support the conclusion that opioids can inhibit GABAergic interneurons in the CNS and can result in disinhibition of neuronal pathways, including pyramidal neurons in the hippocampus (Zieglgansberger et al., 1979). In summary, the nonclinical data support the conclusion that opioids can result in seizures and neurological tissue damage at high doses under the conditions tested.

2.1.2 Human seizure data

Given early nonclinical reports suggesting the potential for seizures and neuropathology in animals it is not surprising that in the early 1980s, several publications investigated the potential for fentanyl to induced abnormal EEG in humans. In what appears to be the first study of anesthetic doses of fentanyl in humans that included EEG recordings, Sebel and colleagues noted that anesthesia induced by fentanyl doses of 30 to 70 μg/kg over 2 min for cardiac surgery resulted in EEG activity consistent with high voltage slow delta waves with no evidence of seizures. They did note that in some patients, the EEG demonstrated isolated sharp wave activity localized to the frontotemporal region, but these did not generalize. Thus, the EEG appearance after fentanyl in their subjects was primarily one of depression. Doses higher than 70 μg/kg were not studied and the authors noted that further investigations should be completed given the published nonclinical data (Sebel et al., 1981).

However, there were also several reports of fentanyl induced convulsions that were published in the clinical literature at this time. For example, Rao et al. reported five cases of convulsions during rapid intravenous administration of clinically relevant doses of fentanyl in patients without any history of seizure disorders (Rao et al., 1982). Safwat and Daniel reported a "grand mal seizure" in a 79 kg patient administered 200 μg of fentanyl in divided doses (Safwat and Daniel, 1983). Hoien subsequently reported a case of "grand mal seizure" in a 17-year-old male who was administered 100 μg of fentanyl for an open reduction of an ankle fracture (Hoien, 1984). In contrast, Murkin and colleagues reported no evidence of abnormal EEG/seizures during anesthesia induction with high-dose fentanyl and noted that the earlier reports did not include any EEG data to substantiate a claim that a seizure actually took place. They report their observations in a small number of human patients following doses of 50 μg/kg ($n = 2$) and 150 μg/kg ($n = 4$) administered over 20 and 60 s, respectively. Peak fentanyl plasma concentrations ranged from 625 to 1750 ng/mL immediately after the 150 μg/kg injection and rapidly declined within 3 min after the injection. Two additional patients were treated with increasing doses of fentanyl intended to maintain plasma fentanyl concentrations up to 625 ng/mL for 15 min. The EEG only showed diffuse slow-wave EEG activity consistent with fentanyl anesthesia. They did note that truncal rigidity was observed in their studies, which has been associated with some motor activity (e.g., ocular movements, wrist or arm flexion), but there was no abnormal EEG activity (Murkin et al., 1984).

Scott and colleagues provided further insights into the potential for fentanyl to induce seizures in their publication of a case report of an 81-kg, 73-year-old male who underwent a surgical procedure that employed fentanyl infusions while having EEG monitoring as part of a clinical study. He was administered fentanyl via intravenous infusion at a rate of 150 μg/min and at approximately 3.5 min after the infusion, he had repeated jerking movements of his right hand and arm. Movements were controlled by succinylcholine administration (skeletal muscle relaxant). The EEG recordings, reviewed by a neurologist, revealed no evidence of seizures during the movements or after succinylcholine administration. At the time of the movements, fentanyl plasma concentrations were 25.3 ng/mL and peaked at 32.0 ng/mL. There was no evidence of neurological injury upon follow up examination by a neurologist. The authors attributed the movements to myoclonus or exaggerations of fentanyl-induced muscle rigidity and that the previous reports of convulsions may not be true seizures (Scott and Sarnquist, 1985).

Smith and colleagues conducted a retrospective analysis of 127 EEG recordings of patients obtained during anesthesia induction with fentanyl, sufentanil, or alfentanil. Of the 127 patients, 93 were determined to have manifested mild to intense muscle rigidity as recorded by electromyographic monitoring (EMG). Of those, 46 were classified as intense and associated with movements that could be interpreted as seizures. Three independent readers examined the EEG without knowledge of the clinical conditions of the patients. They concluded that there was no evidence of epileptiform EEG activity. Random or periodic sharp wave patterns were artifactual and able to be attributed to ECG, infusion pumps, electrode or cable movements, interference, or head movements. Onset of muscle rigidity was generally "explosive in onset" and appeared as flexion of upper extremities, extension lower extremities, immobilization of the head and neck, chin flexion onto the chest, and rigidity of the abdominal and chest wall musculature. Routine administration of muscle relaxants resolved these effects. Smith and colleagues concluded that there was no evidence of seizures in their patients and that the clinical manifestations that gave an appearance of convulsions could be attributed to muscle rigidity or artifact. They do note that they only employed surface electrodes and that they could not provide "incontrovertible evidence" that none of the patients had seizure activity in deeper structures such as the hippocampus or amygdala (Smith et al., 1989).

Sprung and Schedewie published a case report of what appeared to be an apparent focal motor seizure with a Jacksonian March following fentanyl induction. However, they also noted that there was no evidence of seizure in simultaneous EEG recordings and concluded that in their case the events were likely nonepileptic myoclonus. They noted that there are three possible explanations for opioid-related muscle activity. First, the abnormal motor activity could be myoclonus or clonus resulting from blockade of cortical inhibitory pathways resulting in unrepressed excitation of lower brain motor pathways. Second, the abnormal motor activity may be due to exaggerated muscle rigidity which can resemble seizure activity. Third, the abnormal motor activity could be due to subcortical seizures which are difficult to detect via surface electrodes (Sprung and Schedewie, 1992).

Given the animal data suggesting opioids like fentanyl can cause seizures in animals and the lack of clarity regarding whether opioids can also cause seizures in humans, Tempelhoff and colleagues undertook a study to see if fentanyl used as an anesthetic caused alterations on the EEG indicative of seizures in patients with complex partial temporal lobe seizures. Nine patients with bitemporal epidural recording electrode arrays, implanted to determine which temporal lobe gave rise to their seizures, were induced with fentanyl in preparation for secondary craniotomy for anterior temporal lobectomy and evaluated. Each patient consistently demonstrated temporal lobe seizures originating from one temporal lobe. During fentanyl induction that ranged from 17.7 to 35.71 μg/kg, eight of the nine patients demonstrated electrical seizure activity. Interestingly, four of the eight demonstrated fentanyl-related EEG seizure activity in the "healthy" temporal lobe contralateral to the lobe where their seizures were demonstrated to originate. The authors note that they believe this to be the first report of electrical seizure activity following fentanyl anesthesia. The explanation for the observation in some patients of abnormal EEG from the otherwise healthy lobe is not entirely clear. The authors speculated that one explanation is that opioids could in fact induce electrical seizure activity in nonepileptic patients. An alternative explanation is that the otherwise healthy side may have altered opioid receptor density in some patients which may alter the spread of electrical activity (Tempelhoff et al., 1992). Cascino and colleagues reported a similar increased susceptibility to seizures in epileptic patients following alfentanil administration (Cascino et al., 1993).

Kacar Bayram and colleagues reviewed the literature that describe the effects of various anesthetic agents on the quality and reliability of intraoperative electrocortography (ECoG), which employs electrophysiological monitoring of electrical potentials associated with brain activity in the cerebral cortex and is used to locate epileptic foci in humans. Because various anesthetic agents are known to produce nonspecific spike activations or inhibit spike activation, identification of the ideal anesthetic regimen improves the accuracy of ECoG. Based on the authors' review of the literature, fentanyl and alfentanil were not recommended due to their potential to induce neuronal spike patterns. In contrast, remifentanil had no significant effect on spike patterns at low doses (Kacar Bayram et al., 2021). Interestingly, D'Cuhna and colleagues report that use of alfentanil as an anesthetic may prolong therapeutic seizure duration induced by electroconvulsive shock, supporting the possibility that this drug may alter seizure thresholds (D'Cunha et al., 2016) although reduced need for sedation and anesthesia drug requirements may also play a role in this observation. Likewise, remifentanil as the sole anesthetic for therapeutic seizure induction also appeared to reduce the amount of dynamic energy required to induce seizures in patients refractory to ECT with methohexital (Hossain and Sullivan, 2008).

Collectively, the human data suggest that several early reports of seizures following induction with fentanyl and related compounds may not have been actual seizures and the observations may have been due to the known myoclonic effects of fentanyl and related compounds. It may also be possible that attainment of higher brain concentrations of opioids necessary for anesthesia induction quickly bypasses any vulnerable period of increased opioid-related seizure susceptibility or alteration in seizure threshold. Alternatively, any opioid-induced neuroexcitation may be counteracted by other medications used during anesthesia, such as benzodiazepines, with the overall effect being that any opioid-related seizure risk during anesthesia does not appear to result in adverse effects in most patients. The extensive human experience with opioids as anesthetics supports the overall safety of these drugs, when used as indicated.

Although the FDA-approved drug product labels do not suggest that fentanyl or opioids in general cause seizures, the potential for these compounds to increase seizure risk is noted in the warnings and precautions sections of the most recently approved labels. Specifically, many opioid drug product labels in the United States have been updated to Physicians Labeling Rule (PLR) format to include a statement similar to the following:

> 5.X Increased Risk of Seizures in Patients with Seizure Disorders.
>
> The fentanyl in BRANDNAME may increase the frequency of seizures in patients with seizure disorders, and may increase the risk of seizures occurring in other clinical settings associated with seizures. Monitor patients with a history of seizure disorders for worsened seizure control during BRANDNAME therapy.

Nonetheless, the currently used clinical setting is a highly monitored setting where anesthesiologists know patient histories and can rapidly adjust respiration, oxygenation, and cardiovascular parameters as needed. This is obviously not the same scenario when opioids of unknown potency are administered by individuals for recreational purposes.

2.2 Leukoencephalopathy

Toxic leukoencephalopathy has been reported as a rare side effect of opioid exposure. Toxic leukoencephalopathy presents clinically with neurological symptoms including alterations in mental status, visuospatial deficits, even coma and death that are attributed to injury of the cerebral white matter. It can present as either an acute form where symptoms manifest rapidly following insult or a chronic form where symptoms develop slowly. Toxic leukoencephalopathy can be caused by a variety of chemical toxicities including chemotherapeutics, immunosuppressants, antimicrobials, and environmental toxins (Ozutemiz et al., 2019). Acute toxic leukoencephalopathy can present with a sudden increase in intracranial pressure, acute hydrocephalus, and herniation due to cerebellar involvement. Imaging via MRI can assist the diagnosis via demonstration of T2-weighted hyperintensity of the cerebral white matter or via CT scan demonstrating diffuse hypodensity of white matter. Treatment is focused on reducing the increased intracranial pressure to minimize neurological damage which could include steroids, hyperosmolar therapies, vasoactive infusions, and surgical interventions to relieve pressure (Wheaton et al., 2019). Acute hypoxic brain injury generally manifests as acute onset neurological symptoms with persistent effects that can involve gray matter. Delayed post-hypoxic leukoencephalopathy (DPHL) can occur after prolonged periods of hypoxia. Although there may be evidence of recovery from the acute event, onset of neurological symptoms can occur days or even weeks later. MRI findings for DPHL include hyperintense lesions in the pariventricular white matter of both hemispheres and usually involve the frontal lobes. The cerebellum and brainstem are usually not involved (Tormoehlen, 2013). There have been multiple case reports suggesting opioid overdose can be linked to leukoencephalopathies. Tormoehlen noted that in contrast to DPHL, "heroin inhalation leukoencephalopathy" usually presents as a cerebellar syndrome. MRI findings include hyperintense lesions in the cerebellum and posterior cerebral hemispheres and pathology demonstrates intramyelinic vacuolization (Tormoehlen, 2013).

In what appears to be the first report of heroin-associated leukoencephalopathy, in 1982, Wolters and colleagues in the Netherlands reported 47 patients with spongiform leukoencephalopathies and evidence of brown pigment in alveolar macrophages in the lungs. Investigations suggested these cases were attributed to inhalation of "poisonous heroin vapors (pyrolysates)" following the use of black-market heroin. Clinical neurologic signs included motor restlessness, disturbed cerebellar speech, and impaired loud speech in the "initial stage." For those who progressed to an "intermediate stage" more pronounced cerebellar symptoms were noted, including tremors, myoclonic jerks, choreoathetoid movements, and altered reflexes. Eleven of the patients progressed to a "third and terminal stage" which included symptoms of spasms, profuse perspiration, central pyrexia, loss of vision, areflexia and even mutism. These patients all died. There was no evidence of convulsions, hypo- or hypertension, or cardiac arrythmias. All patients demonstrated inconsistent cerebral spinal fluid pleocytosis, depressed EEG, and abnormal brain stem evoked responses. CT scans demonstrated low density cerebellar white matter. Autopsy of 10 of the 11 individuals who died revealed pronounced edema of the entire brain, flattened surface convolutions, and "flabby appearance." Evaluation of brain, brainstem, and spinal cord slices identified tissue edema and "disintegration of parenchyma." Microscopic evaluations identified spongiform degeneration of white matter with "countless vacuoles which coalesced to form larger cavities" and at times focal necrosis in the spongiform white matter. Electron microscopy revealed multivacuolar degeneration of oligodendroglia, swollen mitochondria and distended endoplasmic reticulum in the cytoplasm. Interestingly, samples of heroin were obtained when possible and for "unusual samples" the authors carried out two-week inhalation toxicology studies in rats and rabbits and obtained histopathology data; however, these studies were not able to identify a sample that reproduced the human findings (Wolters et al., 1982). Although these cases clearly suggest inhalation of the heated black-market heroin in these patients can result in very significant adverse effects, the actual neurotoxic causative agent(s) and role, if any, of the opioid, was not clear. The authors did not believe that the effects were related to the heroin itself.

Since that early report, multiple reports of heroin-associated leukoencephalopathies have been reported. For example, Roulet Perez and colleagues reported leukoencephalopathy in a 2-and-a-half-year-old child who may have orally ingested heroin present in the home as a consequence of parenteral use. CT scan showed hypodensities in the cerebellar white matter and MRI scans 3 days later demonstrated bilateral hyperintensities in the same areas. MRI 6 weeks later showed shrinkage of the cerebellar and hemispheric lesions. Again, an impurity in the illicit heroin product was considered the likely cause although in this case it was not attributed to inhalation of heroin pyrolysates (Roulet Perez et al., 1992). Review of the published literature identified numerous heroin-related cases of leukoencephalopathy, particularly after vapor inhalation, a practice commonly referred to as "chasing the dragon" that describes the practice of heating heroin (generally the freebase) over aluminum foil to vaporize the heroin and then inhaling the fumes, which appear like a dragon's tail, through a straw (Ahmad et al., 2021; Buxton et al., 2011; Cordova et al., 2014; Kass-Hout et al., 2011; Keogh et al., 2003; Noori et al., 2020; Singh and Saini, 2015; Zante et al., 2018).

Although the correlation of heroin use and toxic leukoencephalopathy has been repeatedly described in the literature, because the heroin was of unknown purity and reports of leukoencephalopathy are rare given the widespread use of heroin, the causative toxic agent was commonly not thought to be the heroin itself, but rather a toxin formed from the heating or a contaminant present in the heroin. However, more recently encephalopathies associated with what may be pharmaceutical grade opioids have been reported in the published literature.

In 2000, Nanan and colleagues reported a case of a 14-year-old girl who presented with leukoencephalopathy after intentional ingestion of an unknown quantity of morphine sulfate tablets. She was found unresponsive, was intubated and ventilated, and administered naloxone after opioids were detected in her blood. She was in a vegetative state for over a week and developed quadriplegia. Gradual partial recovery occurred over the next year. Plasma morphine levels were extremely high, suggesting a very high dose of morphine was ingested. The initial CT scan was normal, but later scans eventually revealed subtle high signals in the lateral boarder of the fourth ventricle and very high signals in the corpus callosum and globus pallidus. An MRI at 5 weeks demonstrated cerebellar and cerebral atrophy, cystic necrosis of the borderline white matter, and persistent effects in the globus pallidus. As per the authors, delayed hypoxic-ischemic events can result in leukoencephalopathy; however, they concluded that the cerebellar involvement is not consistent with just a delayed hypoxic event and these findings may be more indicative of a toxic leukoencephalopathy (Nanan et al., 2000).

Anselmo and colleagues reported the first case of what appeared to be toxic leukoencephalopathy in a 3-year-old boy who accidentally ingested methadone. The child presented to the emergency department in a coma. CT scans revealed extensive bilateral and symmetrical hypodensity of the cerebellar hemispheres and pons and evidence of acute hydrocephalus. Methadone was detected in the urine. Surgery was performed to reduce the fluid pressure on the brain. By Day 6, MRI confirmed cerebellar lesions and detected bilateral lesions in the hippocampus (Anselmo et al., 2006).

Mills and colleagues reported a 3-year-old who also accidentally ingested methadone and presented with cerebellitis. CT scans revealed swelling and edema of the cerebellum which resulted in compression of the fourth ventricle and hydrocephalus. MRI showed bilateral diffuse cerebellar swelling and edema (Mills et al., 2008). Since these early reports, multiple additional reports of methadone-related encephalopathies have been published (Bileviciute-Ljungar et al., 2014; Hakamifard et al., 2021; Metkees et al., 2015; Salgado et al., 2010). Haghaghi-Morad and colleagues summarize eight cases of methadone-related encephalopathies in patients ranging from 22 years of age to 65 years of age. They broke down the cases into 8 acute-onset encephalopathy (AOE) cases and 2 delayed-onset leukoencephalopathy (DOL) cases based on the delayed timing of presentation of symptoms after methadone intoxication. The goal was to identify and describe any patterns of neurologic deficits following methadone overdose. Retrospective evaluation of MRI data frequently identified bilateral confluent or patchy cerebral white matter hyperintensities and occasional cerebellar abnormalities (Haghighi-Morad et al., 2020).

Morales Odia and colleagues report severe leukoencephalopathy in a 46-year-old man with a history of polysubstance use, alcohol use with delirium tremens and withdrawal seizures, and IV drug use for 10 years. The individual presented in the emergency department with respiratory depression due to acute opioid intoxication after taking considerable amounts of oxycodone pills and other medications for chronic back pain. Based on discussions with the individual and family members, it was believed he ingested 35 tablets of 10-mg oxycodone and five 80-mg oxycodone controlled-release capsules, three carisoprodol 350-mg tablets, and two diazepam pills. Initial CT was normal. He was given respiratory support, treated with naloxone, and later admitted with altered mental status, confusion, and increasing challenges with mobility and coordination. His condition worsened by the second day and he was treated with methadone and lorazepam based on the assumption that some of the symptoms may be related to opioid withdrawal; however, he was later found unresponsive and was again treated with naloxone. Shortly after naloxone, he demonstrated a tonic-clonic seizure which was treated with lorazepam and propofol. He slipped into a comatose state and was moved to ICU. CT scan on Day 3 revealed severe obstructive hydrocephalus. MRI noted edema in both cerebral hemispheres with diffusion positive lesions in both cerebellar hemispheres and globus pallidi. Differential diagnosis ruled out infectious, inflammatory, or coagulation disorders. Although complicated based on the history, the potential contribution of the pharmaceutical grade opioid to this case cannot be ruled out (Morales Odia et al., 2010).

Foy and colleague reported toxic leukoencephalopathy in a 19-month-old girl who presented in the emergency department in a comatose state who was later determined to have a fentanyl patch attached to her back (initial routine drug testing did not detect the presence of any drugs in her system). CT scan showed loss of gray-white matter distinction in the superior cerebellar hemispheres. MRI revealed extensive areas of restricted diffusion throughout the cerebellum bilaterally and patchy areas of restricted diffusion within white matter of both cerebral hemispheres in the frontoparietal and occipital lobes, both caudate nuclei, and the cerebellum that were attributed to hypoxic injury. Following 3 days in intensive care, she demonstrated normal neurological exam except for slightly brisk patellar reflexes and mild ataxia. Evaluation at 3 months was normal, suggesting a full recovery (Foy et al., 2011).

Jones and colleagues reported a case of a 57-year-old female who was found unconscious after allegedly taking 14 tablets of 10-mg oxycodone and not responsive to naloxone infusions. Initial CT scans were normal. She was administered naloxone via infusion for 48 h and a repeated CT scan suggested global hypoxic brain injury. Imaging of the brain over the next 13 days showed attenuation of white matter in the cerebellum and globus pallidus without evidence of an inflammatory, metabolic, or ischemic injury. As per the authors, the pattern of injury that included the globus pallidus was more suggestive of toxic leukoencephalopathy rather than hypoxic brain injury (Jones et al., 2020).

Dinicu and colleagues report a case of an adolescent male who was found unresponsive by a friend approximately 12 h after ingesting what was likely three to four fake oxycodone pills that were apparently laced with fentanyl. MRI imaging found symmetrical diffuse hyperintensities in the corpus callosum and bilateral frontal, parietal, and cerebellum indicative of diffuse white matter injury. Bilateral damage to the basal ganglia was also evident (Dinicu et al., 2020).

Chachkhiani and colleagues reported a case of delayed toxic-hypoxic leukoencephalopathy in a 46-year-old man suffering from respiratory failure and hypotension due to opioid overdose. The patient had a history of substance use, hepatitis C, and mesial temporal lobe epilepsy. On initial examination, brain MRI showed watershed infarcts in the centrum semiovale and EEG showed diffuse theta-wave activity suggested global hypoperfusion. After 5 days he was able to engage in normal activity and was discharged. However, 19 days later he presented with limited responsiveness, abulemia (extreme apathy), a paucity of spontaneous movements, and diffuse hyperreflexia. MRI showed symmetric diffuse white matter hyperintensity with patchy areas of restricted diffusion. The authors speculate that the cause of the delayed white matter deterioration in this case may have been due to a "double hit" initiated by the toxic insult and followed by post-hypoxic progression. However, given the prevalence of opioid overdose and cerebral hypoxia compare to delayed toxic-hypoxic leukoencephalopathy, they also noted that a genetic susceptibility or preexisting condition may contribute to any given case (Chachkhiani et al., 2021).

These cases are challenging to interpret given the potential confounds of polysubstance use, underlying disorders, and variations in the descriptions and diagnoses of the encephalopathies reported. However, collectively, the data suggest that, in addition to heroin pyrolysates, overdose with prescription opioids such as morphine, methadone, and fentanyl may be a contributing factor to these leukoencephalopathies. Obtaining a detailed history of substance use and requesting toxicology screening that includes both routine opioids as well as synthetic opioids should help with the differential diagnosis.

2.3 Opioid-associated amnestic syndrome (OAS)

An amnestic syndrome with evidence of CNS pathology has been reported in recreational opioid users. In 2016, Small and colleagues reported four cases of amnesia with evidence of acute and complete ischemia of the hippocampus in a three-year period in a single tertiary care center. Three of those patients tested positive for opioids and the fourth had history of heroin use. They speculated that the amnesia could be due to toxicity and possibly related to opioid use (Small et al., 2016). In 2017, the same group reported on a cluster of 14 patients in Massachusetts with a history of substance use disorder and an acute anterograde amnestic syndrome that lasted for month or longer. MRI studies noted bilateral hyperintense signals in the hippocampi. Of these patients, 13 tested positive for opioids on routine testing or had a history of opioid use; however, routine testing would not have identified synthetic opioids such as fentanyl (Barash et al., 2017).

In 2018, this same team reported an additional four cases in patients with a history of heroin use and note that initial routine toxicology screens did not detect opioids but that confirmatory testing detected either fentanyl or the metabolite norfentanyl. They note that differential diagnosis of an amnestic syndrome should consider evaluation for synthetic opioids (Barash et al., 2018a). In a subsequent publication, Barash and Kofke built on their initial cases and summarized what was then 18 case reports of sudden-onset amnesia with evidence of acute hippocampal injury reported in Eastern Massachusetts. They reported that 17 of 18 cases tested positive for opioids or had a history of opioid use. Although many of these cases are confounded by other drugs of abuse and use of illicit opioids, they note that the association between opioids and acute amnesia with hippocampal injury deserves further consideration (Barash and Kofke, 2018).

Based on the growing number of cases they have seen in their practices and as reported in the literature, Barash and colleagues proposed a description and definition of "opioid-associated amnestic syndrome" or OAS. As per their proposal, a "confirmed" OAS case definition would include new onset amnesia greater than 24 h in duration, a positive toxicology screen for opioids, and evidence of bilateral hippocampal injury on either CT or MRI. A "probable" OAS case was defined as new onset amnesia greater than 24 h, known history of opioid use, and evidence of bilateral hippocampal injury via CT or MRI. Finally, a "possible" OAS case was defined as a new onset amnesia greater than 24 h and either positive toxicology for opioid, known history of opioid use, or evidence of bilateral hippocampal injury (Barash et al., 2020). The authors concluded that of the 40 reports they identified, 20 were considered "confirmed," 10 were considered "probable," and 10 were considered "possible." Further, they note that although several cases were also linked to cocaine use, because there are reports of cocaine being laced with synthetic fentanyl or fentanyl analogs, it may be possible that some of these individuals may have actually consumed fentanyl or an analog. Because routine testing may not detect these, identification of a possible link to opioid consumption may be missed.

Since this proposed case definition was published several additional independent case reports were also published. Huang and Lukas report a case of a 30-year-old man with a history of polysubstance use, attention-deficit/hyperactivity disorder (ADHD), and multiple suicide attempts presented at the emergency room with anterograde amnesia. MRI examination demonstrated diffusion restriction throughout the bilateral hippocampi. Urine toxicology detected cocaine, benzodiazepines, and opiates (specific testing for fentanyl or other synthetic opioids was not completed). Interestingly, when examined 12 months later, the patient demonstrated no evidence of amnesia, suggesting recovery (Huang and Lukas, 2021).

Ciurylo and Noh reported a case of a 31-year-old male with a history of obesity, ADHD, methamphetamine use, and IV drug use including fentanyl who presented with amnesia greater than 24 h. The individual apparently relapsed on IV fentanyl a week prior to presentation. He tested positive for urine fentanyl and norfentanyl. CT and MRI demonstrated bilateral hippocampal injury and was diagnosed with opioid-associated amnestic syndrome as described by Barash and Kofke (Ciurylo and Noh, 2021).

Walker and colleagues reported a case of a 33-year-old male with a history of polysubstance use who presented in the emergency room after a "heroin" overdose and a history of memory problems over the last several days prior to arriving at the emergency room. Routine urinary drug screen was negative; however, routine screens do not detect synthetic opioids. CT and MRI scans revealed symmetrical bilateral hippocampal edema. After ruling out other potential causes, the authors concluded the patient most likely had opioid-associated amnestic syndrome (Walker et al., 2021).

In response to the 2018 report from Barash et al. in the New England Journal of Medicine, Dutton appropriately pointed out that there are millions of surgical patients a year who are administered fentanyl at levels that exceed the levels most recreational users would achieve and yet there is no signal of similar adverse events being reported from these exposures and suggested that experience supports a conclusion that it is not the fentanyl, but perhaps an impurity causing the OAS (Dutton, 2018). Barash and colleagues responded with two possible explanations for this lack of a signal in the anesthesia literature. One possibility is that the other agents used in conjunction with fentanyl and related substance anesthesia may attenuate the effects on the hippocampus. A second possibility is that patients in an overdose scenario have complications such as hypoxia that may aggravate any fentanyl-related adverse hippocampal effects (Barash et al., 2018b).

In addition to the rodent data generated by Kofke and colleagues suggesting the potential for opioids like fentanyl to result in neuropathology that may contribute to the clinical manifestation of opioid-associated amnestic syndrome, the potential mechanism mediating hippocampal injury and possibly contributing to opioid-associated amnesia was explored in several nonclinical models with varying results. Charchaflieh and colleagues evaluated the effects of fentanyl on anoxic injury to CA1 pyramidal neurons in a rat hippocampal slice model. In their model, pretreatment of hippocampal slices with fentanyl concentrations chosen to mimic the plasma concentrations that have been reported in humans during clinical anesthesia did not alter CA1 population spikes before anoxia or alter recovery after 5, 6, or 7 min of anoxia (Charchaflieh et al., 1998). Similar results were reported in the gerbil model of ischemic neuronal damage (Shiraishi et al., 2011) suggesting that fentanyl pretreatment did not alter vulnerability of the hippocampus to anoxia.

Ammon-Trieber and colleagues studied the effects of opioids on recovery from hypoxia/hypoglycemia in an in vitro rat hippocampal slice model. Their data demonstrated that exposure to opioid antagonists prior to hypoxia/hypoglycemia improved recovery compared to controls suggesting an opioid receptor mediated impact on the tissue response to these adverse conditions. In contrast, exposure to an opioid agonist such as morphine delayed recovery (Ammon-Treiber et al., 2005). Further, their data support the conclusion that exposure to morphine, but not fentanyl or DAMGO, prior to hypoxia/hypoglycemia of the hippocampal tissue significantly impaired recovery during reperfusion as measured via electrophysiological recording of evoked field potentials compared to controls. They speculated that drugs with a greater propensity for receptor internalization, such as fentanyl, were perhaps less neurotoxic than drugs that resulted in less receptor internalization such as morphine (Ammon-Treiber et al., 2007).

Kiyatkin and colleagues employed implanted electrochemical sensors to evaluate brain oxygen levels in the rat model and examined the effects of heroin, fentanyl, morphine, and oxycodone. They reported that although all opioid decreased brain oxygen as a result of respiratory depression, heroin and fentanyl were more potent but also the effects were far more transient than that of morphine and oxycodone given the rapid onset and redistribution of the more lipophilic drugs. They speculated that morphine and oxycodone could actually produce greater neurotoxicity given their longer half-lives in the brain and more prolonged hypoxia under overdose conditions (Kiyatkin, 2019). Vaselbehagh and colleagues reported that intraperitoneal administration of methadone (10 mg/kg) to male mice for 20 consecutive days resulted in increased apoptosis of cells in the CA1 region of the hippocampus (Vaselbehagh et al., 2021). This dose corresponds to a human equivalent dose of 48.7 mg/60 kg person based on body surface area.

Collectively, the clinical case reports suggest a potential role of opioids in the "Opioid-Associated Amnestic Syndrome" described by Barash and Kofke and the adverse effects are consistent with the known vulnerability of the hippocampus to oxygen deprivation. However, further studies are needed to better understand the potential role of therapeutic and/or illicit use of opioids and the impact of opioid overdose in memory impairments.

2.4 Developmental neurotoxicity

Brain development is an extremely complex and adaptive process that involves cellular proliferation and differentiation of embryonic cells to neuronal and glial phenotypes, extensive cellular migration, axon and dendritic outgrowth, programed cell death, synaptogenesis, myelination, and synaptic pruning. The process is driven largely by genetics but also impacted by adaptive or otherwise compensatory changes in response to external influences (Andersen, 2003; Gogtay et al., 2004; Stiles and Jernigan, 2010). Brain development in humans is believed to start in the third week of gestation and extend through late adolescence and early adulthood. The most significant increase in brain mass in humans occurs during the third trimester and the first several years after birth. Historically, this period of rapid increase in brain weight and size has been referred to as the "brain growth spurt." In contrast to humans, rat brain weights increase during the postnatal period and peaks on postnatal day (PND) 7 which is believed to corelate with a human term infant (Dobbing and Sands, 1979). However, brain weight alone does not account for the many different physiological changes taking place during development. Because the brains of different species develop at different times relative to birth, evaluation of reports of developmental neurotoxicity reported in nonclinical models must take into account differential brain development across species. Because most developmental neurotoxicology studies are completed in the rodent model, particularly rats, understanding key developmental benchmarks can help inform cross-species evaluations. Semple and colleagues published a comprehensive comparison of rat to human brain developmental processes that permits a more granular look at cross-species comparisons that was intended to help interpret data on hypoxic-ischemic and traumatic brain injury data (Semple et al., 2013) but can also be used to inform drug-related alterations in brain development.

Modern nonclinical drug development programs include assessments of new drug substances on neuronal development as part of the embryofetal development studies (usually conducted in both rat and rabbit models), in pre- and postnatal development studies (generally the rat model), and, if appropriate, juvenile animal studies. However, even these studies are limited in terms of neurotoxicity endpoints given the complexity of the brain and the many different possible adverse effects that could manifest following insult. For example, pre- and postnatal development studies generally only include testing for functional deficits in offspring such as sexual maturation, reproductive capacity, sensory functions, motor activity, and learning and memory (International Conference on Harmonization of Technical

Requirements for Registration of Pharmaceuticals for Human Use, 2021). Studies to evaluate the impact of a compound on motivation and reward are not included in the standard battery of studies at this time. Because extensive human experience existed for many of the historically used opioid drug substances, early opioid NDA submissions did not necessarily include new toxicology studies that were conducted in accordance with current standards. In fact, for some opioid drug substances, the FDA-approved opioid drug product labels are based entirely on published literature (e.g., morphine and methadone). In contrast, some opioid drug product labels include information from studies completed in accordance with Good Laboratory Practices (GLPs) and with relatively modern study designs (e.g., oxycodone, remifentanil) and other drug product labels include a mixture of GLP studies and published literature (e.g., fentanyl).

Current nonclinical study designs for embryofetal development studies recommend testing doses such that the high dose tested in the pregnant animals should be based on either a toxicity endpoint, saturation of systemic exposure, an exposure margin-based endpoint, a maximum feasible dose, or a limit dose (International Conference on Harmonization of Technical Requirements for Registration of Pharmaceuticals for Human Use, 2021). For opioids, the high-dose selection is almost always based on a dose that produces some degree of minimal maternal toxicity such as body weight loss or exaggerated pharmacodynamics such as respiratory depression or excessive sedation. As such, these studies do not test opioids at levels that mimic those that would be obtained in opioid-tolerant patients or test clinical exposures that may occur in an overdose scenario. In contrast, published studies do not generally include doses that define a no observed adverse effect level (NOAEL). Rather, they tend to push the doses to characterize the toxicologic potential of the drug.

2.4.1 Embryonic brain development

Published nonclinical studies have suggested that very high doses of opioids can lead to alterations in embryonic brain development, including reports of opioid-induced neural tube defects, primarily in hamsters and mice. Harpel and Gautieri (1968) administered a single subcutaneous (SC) injection of 100, 200, 300, 400, or 500 mg/kg of morphine sulfate to pregnant CF-1 mice on Gestation Day (GD) 8 or 9 (Harpel and Gautieri, 1968) and reported a dose related increase incidence of exencephaly at ≥300 mg/kg (24 times the human daily dose of 60 kg). The NOAEL of 100 mg/kg/day corresponds to a human equivalent dose of 487 mg/60 kg person based on body surface area.

Iuliucci and Gautieri administered morphine sulfate to pregnant CF-1 mice as a single SC injection of 200, 300, or 400 mg/kg on GD 8 or 9 (Iuliucci and Gautieri, 1971) and reported an increased incidence of exencephaly in some of the fetuses after the low ($n = 1$) and mid dose ($n = 3$) of morphine but not after the high dose. Maternal toxicity was also evident based on an incidence of convulsions in the pregnant animals. Although morphine treatment produced a few incidences of exencephaly, the finding was not dose-dependent and occurred at a high dose with significant maternal toxicity. The low dose of 200 mg/kg/day corresponds to a human equivalent dose of 976 mg/60 kg person based on body surface area.

Jurand and colleagues reported that administration of methadone doses of 22 to 24 mg/kg to pregnant JBT/Jd mice on GD 9 resulted in an exencephaly incidence of 11% in the embryos. This dose is equivalent to 107 mg methadone administered to a 60 kg person based on body surface area, which is within the highest recommended daily doses for humans for the treatment of opioid addiction (Jurand, 1973). Subsequent studies evaluating the effect of opioids on neural tube development were completed using a mouse embryo model that recapitulated the findings in vivo. These studies concluded that both heroin and methadone produced neural tube defects in mouse embryos that were blocked by the opioid receptor antagonist naloxone (Jurand, 1985). Geber and Schramm reported an increase in neural tube defects following administration of heroin, thebaine, propoxyphene, methadone, morphine, hydromorphone, oxymorphone, and meperidine to pregnant hamsters on GD 8. The doses tested produced significant maternal toxicity including maternal deaths and an increased incidence of cranioschisis in the fetuses. To put these doses into perspective, the lowest dose of morphine tested that produced evidence of exencephaly and cranioschisis was 88 mg/kg which is equivalent to 713.5 mg/60 kg person based on body surface area. The NOAEL of 35 mg/kg is equivalent to 283 mg/60 kg person, higher than any dose likely administered to an opioid-naïve person for pain. In a separate set of animals, the neural tube adverse effects of morphine, hydromorphone, and methadone were antagonized by naloxone suggesting that the effects were mediated by opioid receptors (Geber and Schramm, 1975).

Ciociola and Gautieri (1983) administered morphine sulfate to pregnant mice using subcutaneous infusion of 0.04%, 0.4%, or 4% (2.72, 27.2, or 272 mg/kg, respectively) via miniature infusion pumps implanted on GD 7, 8, 9, or 10. The authors reported exencephaly in some fetuses (Ciociola and Gautieri, 1983), however, significantly increased incidence of exencephaly was observed in only 0.04% morphine implanted on GD 7 group, which overlaps the period of neural tube development in this species. Collectively, these nonclinical reports support the biological plausibility that opioid exposures can lead to neural tube defects under some conditions in some nonclinical species. These data likely encouraged investigations into the potential relationship between opioid use in humans and similar neurodevelopmental effects.

Review of the existing FDA-approved labels reveals data that suggest that when tested in accordance with current high-dose selection recommendations, opioids such as oxycodone did not produce adverse effects on neurological development nor did they identify a risk for neural tube defects in animal models that have been reported at high doses in the literature for many opioids, suggesting a class effect. However, the published studies summarized above suggest that opioids as a class may have the potential to induce neural tube defects at high doses in these nonclinical models by an apparent opioid-receptor mediated mechanism. It is not clear if the effects are direct or indirect, for example, via reduced oxygen or glucose supply to tissues during critical periods of development. Further nonclinical studies may help address the potential mechanism(s) mediating these neural tube effects and may ultimately help inform potential clinical significance of these findings.

In January of 2015, the FDA posted a drug safety communication stating that the Agency reviewed published clinical data on the possible risks of opioid use during pregnancy and the potential risk of neural tube defects (U.S. Food and Drug Administration, 2015). Specifically, the communication refers to two retrospective case control studies that reported an increase in risk for neural tube defects following opioid exposure early in pregnancy (Broussard et al., 2011; Yazdy et al., 2013). Both studies reported that mothers with infants born with neural tube defects were more likely to have reported opioid use in early pregnancy than mothers with infants who were not born with neural tube defects. Although both studies were considered to be generally well-designed, the Agency noted that were still limitations to such data and that further information is needed to determine if opioid use during pregnancy increases the risk of neural tube defects. The Agency noted that it continues to monitor and evaluate the use of pain medicines during pregnancy and will provide updates as new data are obtained.

2.4.2 Neonatal brain development

As discussed earlier, peak brain growth and development occur in the third trimester of pregnancy and over the first several months of life although development continues into young adulthood. Fetal and neonatal exposure to opioids during this time can occur due to a variety of circumstances, including maternal prescription opioid use for analgesia, as part of an anesthetic regimen for critical surgeries than cannot be delayed, via use as a result of opioid use disorder during pregnancy and lactation, or via medication-assisted treatment (MAT) of opioid use disorders under a doctor's supervision. In terms of analgesic exposure during pregnancy, a survey of data from insurance databases between 2005 and 2011 indicated that approximately 14 to 22% of all pregnant women fill at least one prescription for an opioid during pregnancy with regional variation across the states ranging from 9.5 to 41.6% (Bateman et al., 2014; Desai et al., 2014). According to the CDC, data from a 2019 self-reporting survey of pregnant women estimated that 6.6% of respondents used an opioid during pregnancy with 88.8% of those reporting the use for the treatment of pain (Ko et al., 2020). According to the CDC, review of hospital discharge data in the United States between 1999 and 2014 indicated that opioid use disorders in pregnant women at delivery increased from 1.5 cases per 1000 hospital deliveries in 1999 to 6.5 cases per 1000 hospital deliveries in 2014. Based on the 2014 data, the data ranged from 0.7 per 1000 in DC to 48.6 per 1000 in Vermont (Haight et al., 2018).

Published nonclinical literature reports that prenatal and perinatal exposure to opioids produce a variety of changes in the CNS including alteration of a number of neurotransmitter and neuromodulator systems, including opioid and non-opioid systems (Basheer et al., 1992; Bhat et al., 2006a,b; Bianchi et al., 1988; Buisman-Pijlman et al., 2009; Chen et al., 1994; Chiou et al., 2003; De Vries et al., 1991; Di Giulio et al., 1988, 1989; Kunko et al., 1996; McGinty and Ford, 1980; Puppala et al., 2004; Robinson, 2000, 2002; Robinson et al., 1997; Sahraei et al., 2013; Slotkin et al., 1982; Tempel et al., 1994; Vathy et al., 2003; Vestal-Laborde et al., 2014; Wu et al., 2001; Yang et al., 2003, 2006; Zmitrovich et al., 1994), and impairment in various learning and memory tests, motor activity, thermal regulation, nociceptive responses, and sensitivity to drugs, some of which appear to persist into adulthood (Ahmadalipour et al., 2015; Buisman-Pijlman et al., 2009; Chen et al., 2015a,b; Chiou et al., 2003; Enters et al., 1991; Hovious and Peters, 1984, 1985; Klausz et al., 2011; Nasiraei-Moghadam et al., 2013; Niu et al., 2009; Peters, 1977; Vestal-Laborde et al., 2014; Yang et al., 2003, 2006; Zagon et al., 1979; Zagon and McLaughlin, 1978, 1979, 1984; Zmitrovich et al., 1994). The clinical significance of many of these reported nonclinical neurodevelopmental changes are not clear. However, the data indicate that exposure to opioids during development can lead to alterations in the CNS, possibly with lasting effects.

Clinical data demonstrate that opioid use during pregnancy can result in withdrawal symptoms in the newborn after birth. The impact on the child from opioid use disorder in pregnant women was extensively discussed during a 2018 meeting of experts held by the Substance Abuse and Mental Health Services Administration and an excellent overview of this challenging clinical condition has been published (Larson et al., 2019). Historically, the diagnosis of Neonatal Opioid Withdrawal Syndrome (NOWS) is used when the newborn was exposed to opioids. The more expansive diagnosis of Neonatal Abstinence Syndrome (NAS) is used when evidence indicates that the mother was using multiple drugs and the withdrawal signs are attributed to are multiple psychotropic drugs that may or may not include opioids. The support and care of the pregnant women with opioid use disorder is believed to be the best course of action both for the mother and the child and use of MAT is supported by many professional societies and public health organizations (Larson et al., 2019). There are multiple published clinical studies that conclude that prenatal exposure to opioids can result in neurodevelopmental delays in children and a comprehensive review is beyond the scope of this chapter. As summarized by Larson and colleagues, in addition to reports of hydrocephaly and neural tube defects, prenatal opioid exposures have been reported to be associated with smaller fetal head circumference, decreased cerebral volume, deficits in motor development, cognition, executive function, educational performance, and adverse behavioral outcomes later in life. However, the panel noted that interpretation of these reports is complicated by many factors, including, but not necessarily limited to, the identity and extent of substances use by the mother, differences in treatment regimens used to treat NAS/NOWS, genetic and environmental factors, caregiving environment, parenteral behaviors, trauma, and nutritional status. Data are still needed to better understand the impact of the various diverse treatment regimens currently employed to treat NAS and NOWS and how to minimize risks and maximize benefit for both the mother and child.

Prenatal exposure to opioids can also occur outside of use disorders and MAT treatments, such as therapeutic use of opioids for pain control and anesthesia. Untreated pain can also have adverse effects on the neurodevelopment. Nonclinical and clinical studies suggest that neonatal adverse experiences such as repetitive procedural pain (e.g., skin breaks) can contribute to alterations in brain development. In their recent review, Boggini and colleagues summarized published reports of changes in the CNS that correlated with increased incidence of repetitive pain exposures included reports of slower growth of the thalamus, basal ganglia, and total brain volume in the neonate based on imaging studies; and reduced cortical thickness and region-specific brain volumes when measured during school age (Boggini et al., 2021). Studies in neonatal rats demonstrated that repeated exposures to needle pricks from PND 0 to PND 7 decreased pain latencies, increased preference for alcohol, increased latency for exploration and defensive withdrawal behavior, and prolonged memory in a social discrimination study in adult animals (Anand et al., 1999). Studies in animal models suggest that neonatal exposure to a repetitive inflammatory painful stimuli (formalin injections) can also led to increased cell death in the cortex, amygdala, hypothalamus, and thalamus (Anand et al., 2007). Subsequent studies reported that the cell death in this model was correlate with increases in glial fibrillary acidic protein but not caspase-3 suggesting an inflammatory mechanism rather than an apoptotic one and that long-term effects of neonatal exposure to painful stimuli included reduced exploratory behaviors in adult males and learning and memory deficits in adult females, effects that were attenuated by ketamine (Rovnaghi et al., 2008). In contrast, Duhrsen and colleagues reported apoptotic cell death in the frontal and parietal cortex and cingulate gyrus in neonatal rats exposed to repeated formalin injections which were attenuated by morphine (Duhrsen et al., 2013). These observations support the need for adequate analgesia and anesthesia in neonates and very young children. In the case of opioids, data also indicate that opioids play a role in brain development and therefore, studies to understand the potential impact of any opioid medications used during this critical period of brain development also need to be undertaken to understand the overall benefit:risk of interventions.

Interest in the potential impact of anesthetic opioids on brain development during the early brain growth spurt increased when questions about the potential adverse neurodevelopmental impact of anesthetic and sedative drugs that act as NMDA receptor antagonists or GABAergic agonists were raised in the nonclinical literature (Jevtovic-Todorovic et al., 2003; Mellon et al., 2007; Olney et al., 1989). Although opioids were not originally implicated in these findings, they are used in anesthesia and can be either an alternative anesthetic in some cases or can be used to reduce the need for anesthesia in other cases or used to control pain. In addition, methadone not only acts as an agonist at opioid receptors, data suggest that methadone can also antagonize NMDA receptors. The published nonclinical studies examining the impact of opioids during the brain growth spurt have reported mixed results. Several published studies suggested that opioids may not produce the same adverse effects reported for NMDA antagonist and GABAergic agonist anesthetic drugs on brain development.

Rizzi and colleagues used fentanyl as a sham control (30 μg/kg IV slowly over 15 min, followed by 4 h of continuous infusion of 15 μg/kg/h) in their studies on anesthesia-induced developmental effects in the guinea pig model. Prenatal exposure of pregnant dams to an anesthetic cocktail of isoflurane, nitrous oxide, and midazolam increased neuronal apoptosis the fetal brain structures including the retrospenial, parietal, cingulate, occipital, and piriform cortices, the sham control produced a slight increase in apoptosis compared to saline control, although the magnitude of the effect was not statistically different than saline controls (Rizzi et al., 2008). Rizzi and colleagues reported similar findings in the piglet model examining both the parietal and occipital cortices (Rizzi et al., 2010; Sabir et al., 2013). These data supported the conclusion that opioids did not produce a similar neuroapoptotic response as has been reported with other general anesthetic regimens in animal models.

However, more recent studies note that although opioids did not recapitulate the typical effects noted in tissues vulnerable to most general anesthetic drugs, they are not necessarily without risk. Specifically, Sabir and colleagues report that 24 h of intravenous fentanyl under normothermic conditions produced an increase in apoptosis of internal granular cells of the cerebellum in the piglet model. Interestingly, Purkinje cells were not impacted (Sabir et al., 2018).

In contrast to the data reported with fentanyl in neonatal piglets, Bajic and colleagues reported that repeated dosing of neonatal rats with morphine resulted in an increased incidence of apoptosis of neurons compared to saline control treatment. Pups were treated via SC injection with morphine sulfate (10 mg/kg) twice a day from PND 1 to PND 6.5, a treatment regimen that resulted in tolerance to the effects of morphine as measured by antinociceptive responses to the hot-plate and tail withdrawal tests. Twenty-four hours after the last dose, animals were sacrificed on PND 7. CNS tissues (cortex, amygdala, hippocampus, hypothalamus, and periaqueductal gray) were evaluated for apoptosis via caspase-3 coupled with immunohistochemical labeling to differentiate cell types. Morphine treatment increased the incidence of apoptotic neurons only in the cortex and amygdala. There was no clear evidence of apoptosis in microglial cells (Bajic et al., 2013).

Collectively these data support the conclusion that opioid anesthetics may have reduced risk for similar neuroapoptosis compared to NMDA and GABAergic anesthetics; however, opioid anesthetics and analgesics are not necessarily without other long-term effects based on published reports on the perinatal exposure to opioids. Further nonclinical and clinical studies are needed to better understand the impact of opioids on neuronal development in general and specifically when used in the context of necessary medical interventions (e.g., analgesia, anesthesia, NOWS). As with all medications, the benefit:risk of treatment options versus electing to not treat a serious condition must be taken into consideration.

2.4.3 Adolescent brain development

Although many neurodevelopmental changes occur early in life during the early brain growth spurt (Dobbing and Sands, 1979), the periadolescent brain also undergoes considerable changes in structure and function including a second wave of synaptogenesis that involves a second round of overproduction of and elimination of synapses and alterations of receptor densities in tissues. This second wave of synaptogenesis is thought to permit increased plasticity of the brain and to permit the developing brain to adapt to the changing environment and ultimately fine-tune itself to establish an adult phenotype (Blakemore and Choudhury, 2006). The increase in plasticity of the brain during this time period results in adaptation of the nervous system in response to internal and external factors including environmental factors, physiological changes, and experiences, including exposure to stress or various drugs which may impact the maturation of the brain and adult behavioral manifestations, including altered response to drug exposure and vulnerability to substance use disorders (Andersen, 2003; Dow-Edwards et al., 2019).

A full discussion of the reward pathways in the brain is beyond the scope of this chapter and have been extensively reviewed elsewhere (Koob and Volkow, 2010). However, a brief overview of our understanding of the basic opioid reward pathways may help with the discussion of opioid impact on the developing brain that follow. Although a gross oversimplification, drug addiction is an incredibly complex disorder of maladaptive neuroplasticity that occurs in vulnerable individuals and includes adaptive changes following repeated exposure to a drug that alters a person's motivation to seek out the drug including the loss of their ability to control the intake of the drug and upon withdrawal of the drug a negative emotional state that ensues further motivates the individual to seek the drug. As noted by Dr. Nestler, vulnerability is thought to be in part driven by genetic factors and in part by non-genetic factors including environmental exposures and other events that occur during development (Nestler, 2014). The neuronal circuitry believed to be involved in addiction include structures that adapt to the presence of the drug (e.g., ventral tegmental area and ventral striatum), structures that contribute to the withdrawal and negative affect resulting from withdrawal (e.g., extended amygdala), structures that impact the preoccupation with the drug or anticipation of the drug effects including craving (e.g., orbitofrontal cortex-dorsal striatum, prefrontal cortex, basolateral amygdala, hippocampus and insula), and structures involved in loss of inhibitory control (e.g., cingulate gyrus, dorsolateral prefrontal cortex, inferior frontal cortex) (Koob and Volkow, 2010). Historically, two key primary structures in the reward pathways that have been investigated are the ventral tegmental area (VTA) and the nucleus accumbens. The ventral tegmental area contains dopaminergic neurons which project to multiple structures, including, but not limited to, the hippocampus, the amygdala, and the nucleus accumbens. Release of dopamine in the nucleus accumbens is usually associated with reward and positive reinforcement (Di Chiara et al., 2004). Stimulation of mu opioid receptors in the VTA, rostromedial tegmental nucleus, and the ventral pallidum inhibit GABAergic neurotransmission in the VTA resulting in disinhibition of the VTA dopaminergic neurons and release of dopamine in structures innovated by these neurons, including the VTA. The overall impact of opioids on the reward pathways; however, is far more complex and the reader is referred to comprehensive reviews for further details (Fields and Margolis, 2015; Nestler, 2004). However, opioid-induced alterations in these known pathways could impact drug craving, withdrawal, motivation, and loss of inhibitory control over drug seeking behavior. More recent studies have provided data that in addition to neuronal adaptations leading to addiction in an individual suffering from substance use disorder, epigenetic changes have been reported that can impact long-term effects of addiction (Browne et al., 2020; Nestler, 2014).

The adolescent period of development is also characterized by a well-documented increase in impulsivity and risk-taking behavior that can include an increase in drug and alcohol use, criminal behavior, and/or unprotected sex. The propensity to take on added risk for perceived reward during adolescence appears to be in part based in developmental neurobiology (Casey and Jones, 2010).

Galvan and colleagues examined the development of the neural systems believed to be implicated in reward-seeking behavior in a total of 37 subjects aged 7 to 29 years of age via functional magnetic resonance imaging (fMRI). Three groups were compared: children aged 7–11 years, adolescents aged 13–17 years, and adults aged 23–29 years. They hypothesized that the differential risk-taking behaviors may be due to differences in the development of mesolimbic system, known to be implicated in the reward pathways. Subjects were presented with a delayed-response-to-choice task paradigm of three different cues each associated with a specific reward value. The cues (three different pirate images each associated with a specific reward value) were presented on either the left or right side of a monitor screen for 1000 milliseconds (ms) at which time the image was removed. After a 2000 ms delay two reward images (treasure chests with various amounts of coins in them) appeared on the left and right side of the monitor and subjects and were asked to select the left or right side to indicate which side the pirate image was on. After another 2000 ms delay reward feedback was presented in the center of the screen that was associated with the specific pirate image. These events took place within the fMRI and imaging recorded activity in the mesolimbic system, specifically the nucleus accumbens and orbital frontal cortex. Their results indicated that adolescents showed greater activity in the nucleus accumbens in response to large reward compared to children and adults. In contrast, adolescents demonstrated more diffuse activity in the orbital frontal cortex that was more similar to children than adults. The authors interpreted their findings to support the conclusion that the development of nucleus accumbens precedes that of the orbital frontal cortex during adolescence. This neural framework has been speculated to help explain increased risk for addiction in adolescence as well. Specifically, greater response to reward with less "top-down" control via the prefrontal cortex may contribute to poor decision making and increased vulnerability to risk, including addiction (Galvan et al., 2006).

These behavioral and imaging observations are also supported by our understanding of the development of these critical reward system structures using receptor binding studies. For example, Teicher and colleagues characterized the ontogeny of dopamine D1 and D2 receptor expression in the rat striatum and nucleus accumbens. They report that both D1 and D2 receptors in the striatum increase during development and peak around PND 40 and then were clearly reduced over time into adulthood. In contrast, D1 and D2 receptors in the nucleus accumbens also increased up to puberty, but levels were largely maintained into adulthood. These data suggest differential dopaminergic pruning in the nucleus accumbens of adolescent animals compared to adults (Teicher et al., 1995). Tarazi and colleagues demonstrated in rodent models that dopamine D1, D2, and D4 receptors are expressed at higher levels in the nigrostriatal and mesolimbic dopaminergic systems during adolescence compared to adults (Tarazi et al., 1998a,b, 1999).

Differential development of key structures in the central reward pathways could play a role in differential vulnerability to drug misuse at different stages of development. In addition, repeated use of opioids has been shown to produce long lasting changes in these neuronal tissues. Eric Nestler and colleagues provided some of the first evidence that chronic opioid exposure can also result in morphological changes to neurons in tissues in the reward pathways, including the VTA. Noting earlier studies that demonstrate chronic morphine administration results in the induction of tyrosine hydroxylase enzyme (the rate limiting enzyme in dopamine synthesis) (Beitner-Johnson and Nestler, 1991), decreases in levels of neurofilament proteins (Beitner-Johnson et al., 1992), and increased levels of glial fibrillary acidic protein in this region (Beitner-Johnson et al., 1993), they hypothesized that chronic morphine would result in morphologic changes in neurons as well. Their studies demonstrated that 5 days of morphine administration via implanted pellets to rats resulted in selective reduction in the size of the dopaminergic neurons in the VTA and the mean length of the dopamine neuronal processes. The effects were blocked by either naltrexone or microinjection of brain derived nerve growth factor (BDNF) into the VTA (Sklair-Tavron et al., 1996). These adaptive changes in the CNS in response to the presence of opioids can result in long-term effects with functional implications.

A growing body of epidemiology studies support the conclusions that exposure to opioids during adolescence may increase the risk of substance use disorders in young adults. Miech and colleagues analyzed data from 6220 twelfth grade students who were then followed through the age of 23 to determine the incidence of nonmedical use of prescription opioids between the ages of 19 and 23. They report that there was a 33% increase in the risk of future nonmedical use of prescription opioid use if subjects legitimately were exposed to opioids before graduation from high school (relative risk 1.33). The increased risk was present even in subjects without a history of drug use and who had a strong disapproval of illegal drug use when initially surveyed. They do note that the data indicate that legitimate prescription opioid use by 12th grade did not predict frequent opioid misuse. The reasons for the misuse were not necessarily to relieve pain; 69% reported misuse in order to “feel good” or “get high” or to “relieve tension” (Miech et al., 2015).

McCabe and colleagues retrospectively examined data from a cohort of U.S. high school seniors ($n = 4072$) who were surveyed and followed to the age of 35 years to assess the impact of early opioid use on adult use patterns. An initial study reported that the adjusted odds ratio (AOR) associated with substance use disorders in adults age 35 was higher in the sample of high school seniors with a history of non-medical use prescription opioids only (AOR 2.61) both a history of nonmedical use of prescription opioids and a history of medical use of prescription opioids (AOR 1.49) compared to high school seniors who either never had exposure to medical opioids at all or have only a history of medical use of prescription opioids (McCabe et al., 2016). Similar results were reported by this group after prospective examination of a larger cohort ($n = 8373$) of subjects. Interestingly, adolescents who reported medical use without a history of nonmedical use of prescription opioids did not differ from adolescents who had no history of medical use of prescription opioids in terms of risk for future substance use disorders at the age of 35. In contrast, there was increased risk for substance use disorders in adolescents who indicated only nonmedical or medical use of prescription opioids after nonmedical use of prescription opioids (McCabe et al., 2019). This suggests that factors such as the context of the initial experiences and availability may contribute to future risk for substance use disorders. Using a similar study design approach in their expanding cohort of subjects ($n = 11{,}012$), McCabe and colleagues reported an increased odds of "any" heroin use in adulthood in adolescents who reported nonmedical prescription opioid use followed by either medical use or nonmedical prescription opioid misuse. These authors subsequently reported that nearly 1 in 3 adolescents who reported nonmedical prescription opioid use ultimately used heroin in the future by the age of 35. The authors noted that this observation could be due to changes in opioid prescribing (McCabe et al., 2021). Most recently, McCabe and colleagues published reports extending these observations on opioid use patterns during adolescence to predict use patterns from age 35 to age 50. Their analysis of a 26,575 individuals continues to support the conclusion that adolescent prescription opioid misuse increases the risk for opioid misuse later in life (McCabe et al., 2022). Analysis of the motives for misuse of prescription opioids does appear to change with time from predominantly "get high" in younger adults (18–25 years of age) to predominantly "relieve pain" in older adults (Schepis et al., 2020).

Adolescent brain exposure to opioids have been reported to alter neuronal function in adult animals in multiple nonclinical studies. Schwarz and

Bilbo examined the effect of adolescent morphine exposure on subsequent risk of addiction in a rat model. Sprague Dawley rats were treated with morphine sulfate BID for 5 days from PND 37 to 42 (adolescent) or 54 to 58 days (adults) and drug preference for morphine was evaluated using the conditioned place preference (CPP) model. After the initial CPP for morphine was established, both adolescent and adult rats underwent an extinction period and then were retested for drug-induced reinstatement. Adolescent exposed rats consistently demonstrated reinstatement of CPP to morphine as adults. In contrast, adult rats that demonstrated CPP to morphine inconsistently showed reinstatement of morphine CPP with about half of the rats showing evidence of reinstatement and the other half of the rats not showing reinstatement (Schwarz and Bilbo, 2013). Interestingly, this research team earlier reported that morphine reinstatement in the rat correlates with an increase in glia cells cytokine and chemokine expression in the nucleus accumbens core (NAc) and that attenuation of morphine induced glial cell activation in this structure blocks morphine induced reinstatement in the CPP model (Schwarz et al., 2011). Likewise, Schwarz and Bilbo report that pretreatment of adolescent rats with ibudalast, a glial cell inhibitor, prior to morphine-induced CPP blocks the increase in inflammatory mediator expression (TLR4 mRNA expression) in the NAc and also blocks morphine reinstatement in adult rats. In contrast, morphine-induced CPP in adult rats was not associated with altered TLR4 expression in the NAc (Schwarz and Bilbo, 2013). These data suggest a potential role of the neuroinflammation as a potential mechanism of substance use disorders in adolescence (Guerri and Pascual, 2019).

Sanchez and colleagues also examined the potential impact of adolescent opioid exposure on reward pathways in adults. These researchers treated adolescent and adult mice with oxycodone for 28 days and then evaluated their response to morphine in the CPP paradigm after a 28-day opioid-free period. Adolescent mice exposed to oxycodone demonstrated a three-fold increase in the response to morphine CPP as adults. In contrast adult mice exposed to oxycodone via this paradigm did not demonstrate altered morphine CPP when tested after a similar opioid-free period. Interestingly, adolescent oxycodone exposed animals also demonstrated reduced expression of dopamine D1 receptor expression in the nucleus accumbens and dopamine transporter (DAT) expression in the VTA as adults but the same exposure as adults did not result in alteration of these parameters. In contrast to the CPP response, there was no impact of the adolescent oxycodone exposure on other behavioral endpoints, including an open field test, a marble burying, locomotor activity, or time spent immobile in the forced swim test, although oxycodone during adolescence did result in a blunted release of corticosterone in response to the forced swim test compared to controls. These data suggest that adolescent oxycodone exposure results in long-lasting changes in the reward pathway (Sanchez et al., 2016).

Moazen administered morphine BID (10 AM and 10 PM) for 5 days (4 mg/kg Days 1 and 2, 8 mg/kg Days 3–5) to adult (PND 104), young adult (PND 60), and adolescent (PND 45) rats and evaluated impulsivity in the 5-Choice Serial-Reaction Time Test (5-CSRTT). The slightly older age used for adolescent animals was required for training in the battery which started at PND 21 (time of weaning). In this model, food delivery is dependent on waiting until a light flashes brief indicating which one of five locations contains the food. Impulsivity was measured by the number of premature responses to the brief light flash stimulus used to signal that a food pellet was available. The authors report that morphine exposure increased motor impulsivity in all animals on Day 6, but increased impulsivity on Day 30 was only noted in adolescent treated animals, suggesting that adolescent exposure to morphine increased impulsivity for longer periods of time than in adult animals (Moazen et al., 2018).

Collectively, these data suggest that exposure to opioid during adolescence can have a profound impact on brain development and increase the risk for vulnerability to addiction later in life. It should be emphasized; however, that the nonclinical studies are models that do not mimic the approved clinical use of these drugs for the treatment of painful conditions, as an anesthetic, or in conjunction with anesthesia. However, the results may well be more applicable to the scenario of misuse of these compounds for recreational purposes. Nonetheless, the data from human epidemiological studies suggest that the manner in human adolescents are first exposed to these drugs may have a significant impact on their future likeability and misuse.

2.5 Evidence for transgenerational neurotoxicity

The impact of the opioid epidemic has been considerable; however, there are concerning reports that suggest that exposure to opioids can have lasting impact on future generations (Gilardi et al., 2018; Vassoler et al., 2014), suggesting that the overall impact of the epidemic may not yet be fully realized. Byrnes treated 30-day-old female rats with morphine for 20 days followed by a 10-day treatment-free period prior to mating and testing the effects on the adult offspring. She reported that adult female offspring of females exposed to morphine during puberty displayed increase in anxiety behaviors (decreased exploration of novel environments, reduced time spent in open arms of elevated plus maze test) and more rapid induction of morphine sensitization. Male adult offspring demonstrated enhanced expression of morphine sensitization. The author concluded that morphine exposure during adolescence resulted in transgenerational effects including response to opioids (Byrnes, 2005). A subsequent study also reported that male adult offspring of females treated with morphine during puberty demonstrated increased response to morphine and more rapid tolerance to morphine antinociception in the hot plate test (Byrnes et al., 2011). This group also reported an increased functional response to dopamine agonists and increased expression of dopamine D2 receptor and kappa opioid receptor gene expression in the nucleus accumbens of adult F1 and F2 generation offspring of females exposed to opioids during puberty that mated 3 weeks after last opioid administration (Byrnes et al., 2013). Using this same paradigm, these researchers reported a significant increase in the conditioned place preference response to low doses of morphine, significantly increased expression of mu opioid receptors in the nucleus accumbens, and decreased expression of mu opioid receptors in the ventral tegmental area of adult offspring (Vassoler et al., 2016). Subsequent studies demonstrated that adult F1 offspring of morphine treated adolescent females self-administered less morphine and a blunted reinstatement behavioral response suggesting an attenuation of relapse behavior than their control counterparts with no change in the learning of the task. Similar although less pronounced effects were noted in the F2 offspring. The authors state that decreased responding to morphine can be interpreted as a higher drug sensitivity or a decrease in the motivation for the drug. The authors speculate that although these alterations may be viewed as protective, they may also be indicative of a blunted reward system or even a depressive-type phenotype which can alter risk for addictive behaviors (Vassoler et al., 2017). This is further supported by subsequent studies reporting an increase in response to the rewarding effects of cocaine in adult offspring of adolescent females treated with morphine (Vassoler et al., 2019) and decreased anxiety-like behavior in F1 and F2 female offspring (Vassoler and Byrnes, 2021).

Studies in humans to evaluate the potential for transgenerational effects of any drug, chemical, or environmental exposure are extremely challenging given the many potential confounding variables that could impact overall conclusions of causality. However, collectively, these nonclinical data suggest the need for further studies of the potential impact of opioids on neuronal development and specifically studies examining potential epigenetic changes that could contribute to transgenerational effects that could impact public health. If clinical data are obtained that support concern for transgenerational effects, such knowledge could be leveraged by medical professionals to provide even greater personalized care of patients with a family history of substance use disorders.

3. Conclusions

Opioids have had an enormous impact on the overall health and welfare of humans. There is no class of drugs more efficacious to treat pain than opioids. Likewise, the invention of fentanyl and fentanyl analogs has had a tremendous positive impact of the field of anesthesiology. When used in accordance with the FDA-approved drug product labeling, opioids provide a critical need for patients and contribute to an overall improvement of life and wellbeing. However, like most drug products, misuse of opioids can have profound adverse effects. Aside from the known risk of death from opioid overdoses, which have increased in recent years as fentanyl and fentanyl analogs were introduced into the illicit opioid marketplace, those who survive an overdose may still have profound adverse effects, at least in part due to the decreased oxygenation of vulnerable neurological tissues. This appears to be a contribution factor in reports of opioid-related leukoencephalopathy and perhaps the recently described opioid-associated amnestic syndrome. However, even in the absence of significant opioid-related neurological injury in adults, data suggest that opioid exposure during critical windows of development can have adverse consequences on brain development.

Studies in animals that suggest that opioid exposure during adolescence may have multigenerational effects warrant further investigations including studies to ascertain the potential clinical significance of these findings. Although many of these adverse effects of opioids summarized in this chapter are likely limited to situations of opioid overdose, further investigations into the role of opioids on overall brain function and development are warranted to better understand the potential translatability of the nonclinical findings to the clinical setting. In addition, given the significant increase in fentanyl analogs in illicit substances, further studies on the effects of these compounds are also warranted.

References

Ahmad, S., Whiting, A., Song, D., Reyes, J.V., Ireifej, B., Almas, T., Lieber, J.J., 2021. Chasing the dragon and stumbling upon an octopus: A case of heroin-induced leukoencephalopathy and reverse takotsubo cardiomyopathy. Ann. Med. Surg. (Lond) 69, 102797.

Ahmadalipour, A., Sadeghzadeh, J., Vafaei, A.A., Bandegi, A.R., Mohammadkhani, R., Rashidy-Pour, A., 2015. Effects of environmental enrichment on behavioral deficits and alterations in hippocampal BDNF induced by prenatal exposure to morphine in juvenile rats. Neuroscience 305, 372–383.

Ammon-Treiber, S., Stolze, D., Schroder, H., Loh, H., Hollt, V., 2005. Effects of opioid antagonists and morphine in a hippocampal hypoxia/hypoglycemia model. Neuropharmacology 49, 1160–1169.

Ammon-Treiber, S., Stolze, D., Hollt, V., 2007. Differential effects of mu-opioid receptor agonists in a hippocampal hypoxia/hypoglycemia model. Brain Res. 1183, 60–65.

Anand, K.J., Coskun, V., Thrivikraman, K.V., Nemeroff, C.B., Plotsky, P.M., 1999. Long-term behavioral effects of repetitive pain in neonatal rat pups. Physiol. Behav. 66, 627–637.

Anand, K.J., Garg, S., Rovnaghi, C.R., Narsinghani, U., Bhutta, A.T., Hall, R.W., 2007. Ketamine reduces the cell death following inflammatory pain in newborn rat brain. Pediatr. Res. 62.

Andersen, S.L., 2003. Trajectories of brain development: point of vulnerability or window of opportunity? Neurosci. Biobehav. Rev. 27, 3–18.

Anselmo, M., Campos Rainho, A., Do Carmo Vale, M., Estrada, J., Valente, R., Correia, M., Vieira, J.P., Barata, D., 2006. Methadone intoxication in a child: toxic encephalopathy? J. Child Neurol. 21, 618–620.

Armenian, P., Vo, K.T., Barr-Walker, J., Lynch, K.L., 2018. Fentanyl, fentanyl analogs and novel synthetic opioids: A comprehensive review. Neuropharmacology 134, 121–132.

Bajic, D., Commons, K.G., Soriano, S.G., 2013. Morphine-enhanced apoptosis in selective brain regions of neonatal rats. Int. J. Dev. Neurosci. 31, 258–266.

Barash, J.A., Kofke, W.A., 2018. Connecting the dots: an association between opioids and acute hippocampal injury. Neurocase 24, 124–131.

Barash, J.A., Somerville, N., Demaria Jr., A., 2017. Cluster of an unusual amnestic syndrome - Massachusetts, 2012-2016. MMWR Morb. Mortal. Wkly Rep. 66, 76–79.

Barash, J.A., Ganetsky, M., Boyle, K.L., Raman, V., Toce, M.S., Kaplan, S., Lev, M.H., Worth, J.L., Demaria Jr., A., 2018a. Acute amnestic syndrome associated with fentanyl overdose. N. Engl. J. Med. 378, 1157–1158.

Barash, J.A., Ganetsky, M., Demaria, A., Jr., 2018b. More on acute amnestic syndrome associated with fentanyl overdose. N. Engl. J. Med. 378, 2247–2248.

Barash, J.A., Whitledge, J., Watson, C.J., Boyle, K., Lim, C., Lev, M.H., Demaria, A., Jr., Ganetsky, M., 2020. Opioid-associated amnestic syndrome: description of the syndrome and validation of a proposed definition. J. Neurol. Sci. 417, 117048.

Basheer, R., Yang, J., Tempel, A., 1992. Chronic prenatal morphine treatment decreases G alpha s mRNA levels in neonatal frontal cortex. Brain Res. Dev. Brain Res. 70, 145–148.

Bateman, B.T., Hernandez-Diaz, S., Rathmell, J.P., Seeger, J.D., Doherty, M., Fischer, M.A., Huybrechts, K.F., 2014. Patterns of opioid utilization in pregnancy in a large cohort of commercial insurance beneficiaries in the United States. Anesthesiology 120, 1216–1224.

Beitner-Johnson, D., Nestler, E.J., 1991. Morphine and cocaine exert common chronic actions on tyrosine hydroxylase in dopaminergic brain reward regions. J. Neurochem. 57, 344–347.

Beitner-Johnson, D., Guitart, X., Nestler, E.J., 1992. Neurofilament proteins and the mesolimbic dopamine system: common regulation by chronic morphine and chronic cocaine in the rat ventral tegmental area. J. Neurosci. 12, 2165–2176.
Beitner-Johnson, D., Guitart, X., Nestler, E.J., 1993. Glial fibrillary acidic protein and the mesolimbic dopamine system: regulation by chronic morphine and Lewis-Fischer strain differences in the rat ventral tegmental area. J. Neurochem. 61, 1766–1773.
Bhat, R., Chari, G., Rao, R., 2006a. Effects of prenatal cocaine, morphine, or both on postnatal opioid (mu) receptor development. Life Sci. 78, 1478–1482.
Bhat, R., Chari, G., Rao, R., Wirtshafter, D., 2006b. Prenatal cocaine and morphine alter brain cyclin-dependent kinase 5 (Cdk5) activity in rat pups. Neurotoxicol. Teratol. 28, 625–628.
Bianchi, M., Marini, A., Sacerdote, P., Cocco, E., Brini, A., Panerai, A.E., 1988. Effect of chronic morphine on plasma and brain beta endorphin and methionine enkephalin in pregnant rats and in their fetuses or newborn. Neuroendocrinology 47, 89–94.
Bileviciute-Ljungar, I., Haglund, V., Carlsson, J., Von Heijne, A., 2014. Clinical and radiological findings in methadone-induced delayed leukoencephalopathy. J. Rehabil. Med. 46, 828–830.
Blakemore, S.J., Choudhury, S., 2006. Development of the adolescent brain: implications for executive function and social cognition. J. Child Psychol. Psychiatry 47, 296–312.
Boggini, T., Pozzoli, S., Schiavolin, P., Erario, R., Mosca, F., Brambilla, P., Fumagalli, M., 2021. Cumulative procedural pain and brain development in very preterm infants: A systematic review of clinical and preclinical studies. Neurosci. Biobehav. Rev. 123, 320–336.
Broussard, C.S., Rasmussen, S.A., Reefhuis, J., Friedman, J.M., Jann, M.W., Riehle-Colarusso, T., Honein, M.A., 2011. Maternal treatment with opioid analgesics and risk for birth defects. Am. J. Obstet. Gynecol. 204, 314.e1–11.
Browne, C.J., Godino, A., Salery, M., Nestler, E.J., 2020. Epigenetic mechanisms of opioid addiction. Biol. Psychiatry 87, 22–33.
Buisman-Pijlman, F.T., Gerrits, M.A., Van Ree, J.M., 2009. Increased opioid release in specific brain areas in animals exposed to prenatal morphine and emotional stress later in life. Neuroscience 159, 405–413.
Burns, G., Derienz, R.T., Baker, D.D., Casavant, M., Spiller, H.A., 2016. Could chest wall rigidity be a factor in rapid death from illicit fentanyl abuse? Clin. Toxicol. (Phila.) 54, 420–423.
Buxton, J.A., Sebastian, R., Clearsky, L., Angus, N., Shah, L., Lem, M., Spacey, S.D., 2011. Chasing the dragon - characterizing cases of leukoencephalopathy associated with heroin inhalation in British Columbia. Harm Reduct. J. 8, 3.
Byrnes, E.M., 2005. Transgenerational consequences of adolescent morphine exposure in female rats: effects on anxiety-like behaviors and morphine sensitization in adult offspring. Psychopharmacology (Berl) 182, 537–544.
Byrnes, J.J., Babb, J.A., Scanlan, V.F., Byrnes, E.M., 2011. Adolescent opioid exposure in female rats: transgenerational effects on morphine analgesia and anxiety-like behavior in adult offspring. Behav. Brain Res. 218, 200–205.
Byrnes, J.J., Johnson, N.L., Carini, L.M., Byrnes, E.M., 2013. Multigenerational effects of adolescent morphine exposure on dopamine D2 receptor function. Psychopharmacology (Berl) 227, 263–272.
Carlsson, C., Smith, D.S., Keykhah, M.M., Englebach, I., Harp, J.R., 1982. The effects of high-dose fentanyl on cerebral circulation and metabolism in rats. Anesthesiology 57, 375–380.
Cascino, G.D., So, E.L., Sharbrough, F.W., Strelow, D., Lagerlund, T.D., Milde, L.N., O'Brien, P.C., 1993. Alfentanil-induced epileptiform activity in patients with partial epilepsy. J. Clin. Neurophysiol. 10, 520–525.

Casey, B.J., Jones, R.M., 2010. Neurobiology of the adolescent brain and behavior: implications for substance use disorders. J. Am. Acad. Child Adolesc. Psychiatry 49, 1189–1201, (quiz 1285).
Centers For Disease Control And Prevention, 2011. Vital signs: overdoses of prescription opioid pain relievers—United States, 1999–2008. MMWR Morb. Mortal. Wkly Rep. 60, 1487–1492.
Chachkhiani, D., Chimakurthy, A.K., Verdecie, O., Goyne, C.T., Mader, E.C., Jr., 2021. Delayed toxic-hypoxic leukoencephalopathy as sequela of opioid overdose and cerebral hypoxia-ischemia. Cureus 13, e20271.
Charchaflieh, J., Cottrell, J.E., Kass, I.S., 1998. The effect of fentanyl on electrophysiologic recovery of CA 1 pyramidal cells from anoxia in the rat hippocampal slice. Anesth. Analg. 87, 68–71.
Chen, J., Ravis, W.R., Walters, D.E., 1994. Perinatal exposure to morphine alters the density, but not the sensitivity, of dopamine receptors in the 10-day-old rat brain. Res. Commun. Alcohol Subst. Abuse 15, 113–123.
Chen, H.H., Chiang, Y.C., Yuan, Z.F., Kuo, C.C., Lai, M.D., Hung, T.W., Ho, I.K., Chen, S.T., 2015a. Buprenorphine, methadone, and morphine treatment during pregnancy: behavioral effects on the offspring in rats. Neuropsychiatr. Dis. Treat. 11, 609–618.
Chen, H.H., Chiang, Y.C., Yuan, Z.F., Kuo, C.C., Lai, M.D., Hung, T.W., Ho, I.K., Chen, S.T., 2015b. Buprenorphine, methadone, and morphine treatment during pregnancy: behavioral effects on the offspring in rats. Neuropsychiatr. Dis. Treat. 11, 609–618.
Chiou, L.C., Yeh, G.C., Fan, S.H., How, C.H., Chuang, K.C., Tao, P.L., 2003. Prenatal morphine exposure decreases analgesia but not K+ channel activation. Neuroreport 14, 239–242.
Ciociola, A.A., Gautieri, R.F., 1983. Evaluation of the teratogenicity of morphine sulfate administered via a miniature implantable pump. J. Pharm. Sci. 72, 742–745.
Ciurylo, W., Noh, E., 2021. Opioid-associated amnestic syndrome. Cureus 13, e16714.
Cordova, J.P., Balan, S., Romero, J., Korniyenko, A., Alviar, C.L., Paniz-Mondolfi, A., Jean, R., 2014. 'Chasing the dragon': new knowledge for an old practice. Am. J. Ther. 21, 52–55.
Corssen, G., Domino, E.F., Sweet, R.B., 1964. Neuroleptanalgesia and anesthesia. Anesth. Analg. 43, 748–763.
Cumming, P., Marton, J., Lilius, T.O., Olberg, D.E., Rominger, A., 2019. A survey of molecular imaging of opioid receptors. Molecules 24.
Davis, M.P., Behm, B., 2020. Reasons to avoid fentanyl. Ann. Palliat. Med. 9, 611–624.
D'cunha, C., Plakiotis, C., O'connor, D.W., 2016. Alfentanil anesthetic augmentation lengthens seizure duration in electroconvulsive therapy with older people. Int. Psychogeriatr. 28, 1051–1052.
De Castro, J., Andrieu, S., Clerckx, A., 1979a. The use of almitrine following potentialized analgesic. Anesthesia based upon alfentanil and flunitrazepam or etomidate. Acta Anaesthesiol. Belg. 30 Suppl, 135–149.
De Castro, J., Van De Water, A., Wouters, L., Xhonneux, R., Reneman, R., Kay, B., 1979b. Comparative study of cardiovascular, neurological and metabolic side effects of 8 narcotics in dogs. Pethidine, piritramide, morphine, phenoperidine, fentanyl, R 39 209, sufentanil, R 34 995. I. Comparative study on the acute toxicity and hemodynamic effects of the narcotics in high and massive doses in curarised and mechanically ventilated dogs. Acta Anaesthesiol. Belg. 30, 5–54.
De Castro, J., Van De Water, A., Wouters, L., Xhonneux, R., Reneman, R., Kay, B., 1979c. Comparative study of cardiovascular, neurological and metabolic side effects of 8 narcotics in dogs. Pethidine, piritramide, morphine, phenoperidine, fentanyl, R 39 209,

sufentanil, R 34 995. II. Comparative study on the epileptoid activity of the narcotics used in high and massive doses in curarised and mechanically ventilated dogs. Acta Anaesthesiol. Belg. 30, 55–69.

De Castro, J., Van De Water, A., Wouters, L., Xhonneux, R., Reneman, R., Kay, B., 1979d. Comparative study of cardiovascular, neurological and metabolic side effects of 8 narcotics in dogs. Pethidine, piritramide, morphine, phenoperidine, fentanyl, R 39 209, sufentanil, R 34 995. III. Comparative study of the acute metabolic toxicity of the narcotics used in high and massive doses in curarised and mechanically ventilated dogs. Acta Anaesthesiol. Belg. 30, 71–90.

De Vries, T.J., Van Vliet, B.J., Hogenboom, F., Wardeh, G., Van Der Laan, J.W., Mulder, A.H., Schoffelmeer, A.N., 1991. Effect of chronic prenatal morphine treatment of mu-opioid receptor-regulated adenylate cyclase activity and neurotransmitter release in rat brain slices. Eur. J. Pharmacol. 208, 97–104.

Desai, R.J., Hernandez-Diaz, S., Bateman, B.T., Huybrechts, K.F., 2014. Increase in prescription opioid use during pregnancy among Medicaid-enrolled women. Obstet. Gynecol. 123, 997–1002.

Di Chiara, G., Bassareo, V., Fenu, S., De Luca, M.A., Spina, L., Cadoni, C., Acquas, E., Carboni, E., Valentini, V., Lecca, D., 2004. Dopamine and drug addiction: the nucleus accumbens shell connection. Neuropharmacology 47 (Suppl 1), 227–241.

Di Giulio, A.M., Restani, P., Galli, C.L., Tenconi, B., La Croix, R., Gorio, A., 1988. Modified ontogenesis of enkephalin and substance P containing neurons after perinatal exposure to morphine. Toxicology 49, 197–201.

Di Giulio, A.M., Tenconi, B., Mantegazza, P., Gorio, A., 1989. The development of met-enkephalin innervation in the brain is altered by perinatal exposure to morphine. J. Endocrinol. Invest. 12, 5–8.

Dinicu, A.I., Chaudhari, A., Kayyal, S., 2020. Diffuse subcortical white matter injury and bilateral basal ganglia neuronal loss after acute opioid overdose. Neuroradiol. J. 33, 267–270.

Dobbing, J., Sands, J., 1979. Comparative aspects of the brain growth spurt. Early Hum. Dev. 3, 79–83.

Dow-Edwards, D., Macmaster, F.P., Peterson, B.S., Niesink, R., Andersen, S., Braams, B.R., 2019. Experience during adolescence shapes brain development: from synapses and networks to normal and pathological behavior. Neurotoxicol. Teratol. 76, 106834.

Drug Enforcement Administration, 2018. Schedules of controlled substances: temporary placement of fentanyl-related substances in schedule I. Fed. Regist. 83, 5188–5192.

Duhrsen, L., Simons, S.H., Dzietko, M., Genz, K., Bendix, I., Boos, V., Sifringer, M., Tibboel, D., Felderhoff-Mueser, U., 2013. Effects of repetitive exposure to pain and morphine treatment on the neonatal rat brain. Neonatology 103, 35–43.

Dutton, R.P., 2018. More on acute amnestic syndrome associated with fentanyl overdose. N. Engl. J. Med. 378, 2247.

Enters, E.K., Guo, H.Z., Pandey, U., Ko, D.J., Robinson, S.E., 1991. The effect of prenatal methadone exposure on development and nociception during the early postnatal period of the rat. Neurotoxicol. Teratol. 13, 161–166.

Feldman, P.L., James, M.K., Brackeen, M.F., Bilotta, J.M., Schuster, S.V., Lahey, A.P., Lutz, M.W., Johnson, M.R., Leighton, H.J., 1991. Design, synthesis, and pharmacological evaluation of ultrashort- to long-acting opioid analgetics. J. Med. Chem. 34, 2202–2208.

Fields, H.L., Margolis, E.B., 2015. Understanding opioid reward. Trends Neurosci. 38, 217–225.

Foy, L., Seeyave, D.M., Bradin, S.A., 2011. Toxic leukoencephalopathy due to transdermal fentanyl overdose. Pediatr. Emerg. Care 27, 854–856.

Freeman, J., Ingvar, H., 1967. Effects of fentanyl on cerebral cortical blood flow and EEG in the cat. Acta Anaesthesiol. Scand. 11, 381–391.
Frenk, H., 1983. Pro- and anticonvulsant actions of morphine and the endogenous opioids: involvement and and interactions of multiple opioid and non-opioid systems. Brain Res. Rev. 6, 197–210.
Frenk, H., McCarty, B.C., Liebeskind, J.C., 1978a. Different brain areas mediate the analgesic and epileptic properties of enkephalins. Science 200, 335–337.
Frenk, H., Urca, G., Liebeskind, J.C., 1978b. Epileptic properties of leucine- and methionine-enkephalin: comparison with morphine and reversibility by naloxone. Brain Res. 147, 327–337.
Frenk, H., Liban, A., Balamuth, R., Urca, G., 1982. Opiate and non-opiate aspects of morphine induced seizures. Brain Res. 253, 253–261.
Galvan, A., Hare, T.A., Parra, C.E., Penn, J., Voss, H., Glover, G., Casey, B.J., 2006. Earlier development of the accumbens relative to orbitofrontal cortex might underlie risk-taking behavior in adolescents. J. Neurosci. 26, 6885–6892.
Geber, W.F., Schramm, L.C., 1975. Congenital malformations of the central nervous system produced by narcotic analgesics in the hamster. Am. J. Obstet. Gynecol. 123, 705–713.
Gilardi, F., Augsburger, M., Thomas, A., 2018. Will widespread synthetic opioid consumption induce epigenetic consequences in future generations? Front. Pharmacol. 9, 702.
Gladden, R.M., Martinez, P., Seth, P., 2016. Fentanyl law enforcement submissions and increases in synthetic opioid-involved overdose deaths - 27 states, 2013-2014. MMWR Morb. Mortal. Wkly Rep. 65, 837–843.
Gogtay, N., Giedd, J.N., Lusk, L., Hayashi, K.M., Greenstein, D., Vaituzis, A.C., Nugent III, T.F., Herman, D.H., Clasen, L.S., Toga, A.W., Rapoport, J.L., Thompson, P.M., 2004. Dynamic mapping of human cortical development during childhood through early adulthood. Proc. Natl. Acad. Sci. U. S. A. 101, 8174–8179.
Guerri, C., Pascual, M., 2019. Impact of neuroimmune activation induced by alcohol or drug abuse on adolescent brain development. Int. J. Dev. Neurosci. 77, 89–98.
Haghighi-Morad, M., Naseri, Z., Jamshidi, N., Hassanian-Moghaddam, H., Zamani, N., Ahmad-Molaei, L., 2020. Methadone-induced encephalopathy: a case series and literature review. BMC Med. Imaging 20, 6.
Haight, S.C., Ko, J.Y., Tong, V.T., Bohm, M.K., Callaghan, W.M., 2018. Opioid use disorder documented at delivery hospitalization—United States, 1999-2014. MMWR Morb. Mortal. Wkly Rep. 67, 845–849.
Hakamifard, A., Hajiahmadi, S., Khorvash, F., Azish, S., 2021. A case study of methadone-induced delayed post-hypoxic leukoencephalopathy with improvement by antioxidant therapy. North Clin. Istanb 8, 106–108.
Harpel, H.S., Jr., Gautieri, R.F., 1968. Morphine-induced fetal malformations. I. Exencephaly and axial skeletal fusions. J. Pharm. Sci. 57, 1590–1597.
Hoien, A.O., 1984. Another case of grand mal seizure after fentanyl administration. Anesthesiology 60, 387–388.
Hossain, A., Sullivan, P., 2008. The effects of age and sex on electroconvulsive therapy using remifentanil as the sole anesthetic agent. J. ECT 24, 232–235.
Hovious, J.R., Peters, M.A., 1984. Analgesic effect of opiates in offspring of opiate-treated female rats. Pharmacol. Biochem. Behav. 21, 555–559.
Hovious, J.R., Peters, M.A., 1985. Opiate self-administration in adult offspring of methadone-treated female rats. Pharmacol. Biochem. Behav. 22, 949–953.
Huang, D., Lukas, R.V., 2021. Complete bilateral hippocampal diffusion restriction and reversible amnesia following opiate, cocaine, and benzodiazepine abuse. Cureus 13, e12651.
International Conference On Harmonization Of Technical Requirements For Registration Of Pharmaceuticals For Human Use, 2021. Detection of Reproductive and

Developmental Toxicity for Human Pharmaceuticals S5(R3) [Online]. Available: https://database.ich.org/sites/default/files/S5-R3_Step4_Guideline_2020_0218_1.pdf. (Accessed 2/28/2022).
Iuliucci, J.D., Gautieri, R.F., 1971. Morphine-induced fetal malformations II: influence of histamine and diphenhydramine. J. Pharm. Sci. 60, 420–425.
Janssen, P.A., 1962. A review of the chemical features associated with strong morphine-like activity. Br. J. Anaesth. 34, 260–268.
Jevtovic-Todorovic, V., Hartman, R.E., Izumi, Y., Benshoff, N.D., Dikranian, K., Zorumski, C.F., Olney, J.W., Wozniak, D.F., 2003. Early exposure to common anesthetic agents causes widespread neurodegeneration in the developing rat brain and persistent learning deficits. J. Neurosci. 23, 876–882.
Jones, E., Umasankar, U., Mallu, H., Hampton, T., Kulendran, A., Patel, M., 2020. Lesson of the month: oxycodone-induced leukoencephalopathy: a rare diagnosis. Clin. Med. (Lond.) 20, 600–602.
Jurand, A., 1973. Teratogenic activity of methadone hydrochloride in mouse and chick embryos. J. Embryol. Exp. Morphol. 30, 449–458.
Jurand, A., 1985. The interference of naloxone hydrochloride in the teratogenic activity of opiates. Teratology 31, 235–240.
Kacar Bayram, A., Yan, Q., Isitan, C., Rao, S., Spencer, D.D., Alkawadri, R., 2021. Effect of anesthesia on electrocorticography for localization of epileptic focus: literature review and future directions. Epilepsy Behav. 118, 107902.
Kass-Hout, T., Kass-Hout, O., Darkhabani, M.Z., Mokin, M., Mehta, B., Radovic, V., 2011. "Chasing the dragon"—heroin-associated spongiform leukoencephalopathy. J. Med. Toxicol. 7, 240–242.
Kay, B., Pleuvry, B., 1980. Human volunteer studies of Alfentanyl (R39209), a new short-acting narcotic analgesic. Anaesthesia 35, 952–956.
Keogh, C.F., Andrews, G.T., Spacey, S.D., Forkheim, K.E., Graeb, D.A., 2003. Neuroimaging features of heroin inhalation toxicity: "chasing the dragon". AJR Am. J. Roentgenol. 180, 847–850.
Kiyatkin, E.A., 2019. Respiratory depression and brain hypoxia induced by opioid drugs: morphine, oxycodone, heroin, and fentanyl. Neuropharmacology 151, 219–226.
Klausz, B., Pinter, O., Sobor, M., Gyarmati, Z., Furst, Z., Timar, J., Zelena, D., 2011. Changes in adaptability following perinatal morphine exposure in juvenile and adult rats. Eur. J. Pharmacol. 654, 166–172.
Ko, J.Y., D'angelo, D.V., Haight, S.C., Morrow, B., Cox, S., Salvesen Von Essen, B., Strahan, A.E., Harrison, L., Tevendale, H.D., Warner, L., Kroelinger, C.D., Barfield, W.D., 2020. Vital signs: prescription opioid pain reliever use during pregnancy - 34 U.S. jurisdictions, 2019. MMWR Morb. Mortal. Wkly Rep. 69, 897–903.
Kofke, W.A., Garman, R.H., Tom, W.C., Rose, M.E., Hawkins, R.A., 1992. Alfentanil-induced hypermetabolism, seizure, and histopathology in rat brain. Anesth. Analg. 75, 953–964.
Kofke, W.A., Garman, R.H., Janosky, J., Rose, M.E., 1996a. Opioid neurotoxicity: neuropathologic effects in rats of different fentanyl congeners and the effects of hexamethonium-induced normotension. Anesth. Analg. 83, 141–146.
Kofke, W.A., Garman, R.H., Stiller, R.L., Rose, M.E., Garman, R., 1996b. Opioid neurotoxicity: fentanyl dose-response effects in rats. Anesth. Analg. 83, 1298–1306.
Kofke, W.A., Garman, R.H., Garman, R., Rose, M.E., 1999. Opioid neurotoxicity: fentanyl-induced exacerbation of cerebral ischemia in rats. Brain Res. 818, 326–334.
Kofke, W.A., Garman, R.H., Garman, R., Rose, M., 2000. Opioid neurotoxicity: role of neurotransmitter systems. Neurol. Res. 22, 733–737.
Koob, G.F., Volkow, N.D., 2010. Neurocircuitry of addiction. Neuropsychopharmacology 35, 217–238.

Kunko, P.M., Smith, J.A., Wallace, M.J., Maher, J.R., Saady, J.J., Robinson, S.E., 1996. Perinatal methadone exposure produces physical dependence and altered behavioral development in the rat. J. Pharmacol. Exp. Ther. 277, 1344–1351.
Kuroiwa, T., Okeda, R., 1994. Neuropathology of cerebral ischemia and hypoxia: recent advances in experimental studies on its pathogenesis. Pathol. Int. 44, 171–181.
Larson, J.J., Graham, D.L., Singer, L.T., Beckwith, A.M., Terplan, M., Davis, J.M., Martinez, J., Bada, H.S., 2019. Cognitive and behavioral impact on children exposed to opioids during pregnancy. Pediatrics 144.
Maekawa, T., Tommasino, C., Shapiro, H.M., 1984. Local cerebral blood flow with fentanyl-induced seizures. J. Cereb. Blood Flow Metab. 4, 88–95.
McCabe, S.E., Veliz, P., Schulenberg, J.E., 2016. Adolescent context of exposure to prescription opioids and substance use disorder symptoms at age 35: a national longitudinal study. Pain 157, 2173–2178.
McCabe, S.E., Veliz, P.T., Boyd, C.J., Schepis, T.S., McCabe, V.V., Schulenberg, J.E., 2019. A prospective study of nonmedical use of prescription opioids during adolescence and subsequent substance use disorder symptoms in early midlife. Drug Alcohol Depend. 194, 377–385.
McCabe, S.E., Boyd, C.J., Evans-Polce, R.J., McCabe, V.V., Schulenberg, J.E., Veliz, P.T., 2021. Pills to powder: A 17-year transition from prescription opioids to heroin among US adolescents followed into adulthood. J. Addict. Med. 15, 241–244.
McCabe, S.E., Schulenberg, J.E., Schepis, T.S., Evans-Polce, R.J., Wilens, T.E., McCabe, V.V., Veliz, P.T., 2022. Trajectories of prescription drug misuse among US adults from ages 18 to 50 years. JAMA Netw. Open 5, e2141995.
McGinty, J.F., Ford, D.H., 1980. Effects of prenatal methadone on rat brain catecholamines. Dev. Neurosci. 3, 224–234.
Mellon, R.D., Simone, A.F., Rappaport, B.A., 2007. Use of anesthetic agents in neonates and young children. Anesth. Analg. 104, 509–520.
Metkees, M., Meesa, I.R., Srinivasan, A., 2015. Methadone-induced acute toxic leukoencephalopathy. Pediatr. Neurol. 52, 256–257.
Miech, R., Johnston, L., O'Malley, P.M., Keyes, K.M., Heard, K., 2015. Prescription opioids in adolescence and future opioid misuse. Pediatrics 136, e1169–e1177.
Mills, F., MacLennan, S.C., Devile, C.J., Saunders, D.E., 2008. Severe cerebellitis following methadone poisoning. Pediatr. Radiol. 38, 227–229.
Milroy, C.M., Parai, J.L., 2011. The histopathology of drugs of abuse. Histopathology 59, 579–593.
Minkowski, C.P., Epstein, D., Frost, J.J., Gorelick, D.A., 2012. Differential response to IV carfentanil in chronic cocaine users and healthy controls. Addict. Biol. 17, 149–155.
Moazen, P., Azizi, H., Salmanzadeh, H., Semnanian, S., 2018. Adolescent morphine exposure induces immediate and long-term increases in impulsive behavior. Psychopharmacology (Berl) 235, 3423–3434.
Morales Odia, Y., Jinka, M., Ziai, W.C., 2010. Severe leukoencephalopathy following acute oxycodone intoxication. Neurocrit. Care 13, 93–97.
Murkin, J.M., Moldenhauer, C.C., Hug Jr., C.C., Epstein, C.M., 1984. Absence of seizures during induction of anesthesia with high-dose fentanyl. Anesth. Analg. 63, 489–494.
Nanan, R., Von Stockhausen, H.B., Petersen, B., Solymosi, L., Warmuth-Metz, M., 2000. Unusual pattern of leukoencephalopathy after morphine sulphate intoxication. Neuroradiology 42, 845–848.
Nasiraei-Moghadam, S., Sherafat, M.A., Safari, M.S., Moradi, F., Ahmadiani, A., Dargahi, L., 2013. Reversal of prenatal morphine exposure-induced memory deficit in male but not female rats. J. Mol. Neurosci. 50, 58–69.
Nestler, E.J., 2004. Historical review: molecular and cellular mechanisms of opiate and cocaine addiction. Trends Pharmacol. Sci. 25, 210–218.

Nestler, E.J., 2014. Epigenetic mechanisms of drug addiction. Neuropharmacology 76 (Pt B), 259–268.
Niemegeers, C.J., Schellekens, K.H., Van Bever, W.F., Janssen, P.A., 1976. Sufentanil, a very potent and extremely safe intravenous morphine-like compound in mice, rats and dogs. Arzneimittelforschung 26, 1551–1556.
Niu, L., Cao, B., Zhu, H., Mei, B., Wang, M., Yang, Y., Zhou, Y., 2009. Impaired in vivo synaptic plasticity in dentate gyrus and spatial memory in juvenile rats induced by prenatal morphine exposure. Hippocampus 19, 649–657.
Noori, M.A.M., Elkattawy, S., Younes, I., Alyacoub, R., Desai, D., 2020. An atypical presentation of heroin inhalation induced leukoencephalopathy (chasing the dragon). Cureus 12, e11215.
O'Brien, C.P., 2017. Drug use disorders and addiction. In: Brunton, L.L., Hilal-Dandan, R., Knollmann, B.C. (Eds.), Goodman & Gilman's: The Pharmacological Basis of Therapeutics, thirteenth ed. McGraw Hill, New York.
Olney, J.W., Labruyere, J., Price, M.T., 1989. Pathological changes induced in cerebrocortical neurons by phencyclidine and related drugs. Science 244, 1360–1362.
Ozutemiz, C., Roshan, S.K., Kroll, N.J., Benson, J.C., Rykken, J.B., Oswood, M.C., Zhang, L., McKinney, A.M., 2019. Acute toxic leukoencephalopathy: etiologies, imaging findings, and outcomes in 101 patients. AJNR Am. J. Neuroradiol. 40, 267–275.
Pergolizzi Jr., J.V., Webster, L.R., Vortsman, E., Ann Lequang, J., Raffa, R.B., 2021. Wooden chest syndrome: the atypical pharmacology of fentanyl overdose. J. Clin. Pharm. Ther. 46, 1505–1508.
Peters, M.A., 1977. The effect of maternally administered methadone on brain development in the offspring. J. Pharmacol. Exp. Ther. 203, 340–346.
Przewlocka, B., Stala, L., Lason, W., Przewlocki, R., 1983. The effect of various opiate receptor agonists on the seizure threshold in the rat. Is dynorphin an endogenous anticonvulsant? Life Sci. 33 Suppl 1, 595–598.
Puppala, B.L., Matwyshyn, G., Bhalla, S., Gulati, A., 2004. Role of endothelin in neonatal morphine tolerance. J. Cardiovasc. Pharmacol. 44 (Suppl 1), S383–S385.
Ramirez, J.M., Burgraff, N.J., Wei, A.D., Baertsch, N.A., Varga, A.G., Baghdoyan, H.A., Lydic, R., Morris, K.F., Bolser, D.C., Levitt, E.S., 2021. Neuronal mechanisms underlying opioid-induced respiratory depression: our current understanding. J. Neurophysiol. 125, 1899–1919.
Rao, T.L., Mummaneni, N., El-Etr, A.A., 1982. Convulsions: an unusual response to intravenous fentanyl administration. Anesth. Analg. 61, 1020–1021.
Rizzi, S., Carter, L.B., Ori, C., Jevtovic-Todorovic, V., 2008. Clinical anesthesia causes permanent damage to the fetal Guinea pig brain. Brain Pathol. 18, 198–210.
Rizzi, S., Ori, C., Jevtovic-Todorovic, V., 2010. Timing versus duration: determinants of anesthesia-induced developmental apoptosis in the young mammalian brain. Ann. N. Y. Acad. Sci. 1199, 43–51.
Robinson, S.E., 2000. Effect of prenatal opioid exposure on cholinergic development. J. Biomed. Sci. 7, 253–257.
Robinson, S.E., 2002. Effects of perinatal buprenorphine and methadone exposures on striatal cholinergic ontogeny. Neurotoxicol. Teratol. 24, 137–142.
Robinson, S.E., Maher, J.R., Wallace, M.J., Kunko, P.M., 1997. Perinatal methadone exposure affects dopamine, norepinephrine, and serotonin in the weanling rat. Neurotoxicol. Teratol. 19, 295–303.
Roulet Perez, E., Maeder, P., Rivier, L., Deonna, T., 1992. Toxic leucoencephalopathy after heroin ingestion in a 21/2-year-old child. Lancet 340, 729.
Rovnaghi, C.R., Garg, S., Hall, R.W., Bhutta, A.T., Anand, K.J., 2008. Ketamine analgesia for inflammatory pain in neonatal rats: a factorial randomized trial examining long-term effects. Behav. Brain Funct. 4, 35.

Rudd, R.A., Paulozzi, L.J., Bauer, M.J., Burleson, R.W., Carlson, R.E., Dao, D., Davis, J.W., Dudek, J., Eichler, B.A., Fernandes, J.C., Fondario, A., Gabella, B., Hume, B., Huntamer, T., Kariisa, M., Largo, T.W., Miles, J., Newmyer, A., Nitcheva, D., Perez, B.E., Proescholdbell, S.K., Sabel, J.C., Skiba, J., Slavova, S., Stone, K., Tharp, J.M., Wendling, T., Wright, D., Zehner, A.M., Centers for Disease Control and Prevention (CDC), 2014. Increases in heroin overdose deaths - 28 states, 2010 to 2012. MMWR Morb. Mortal. Wkly Rep. 63, 849–854.

Sabir, H., Bishop, S., Cohen, N., Maes, E., Liu, X., Dingley, J., Thoresen, M., 2013. Neither xenon nor fentanyl induces neuroapoptosis in the newborn pig brain. Anesthesiology 119, 345–357.

Sabir, H., Dingley, J., Scull-Brown, E., Chakkarapani, E., Thoresen, M., 2018. Fentanyl induces cerebellar internal granular cell layer apoptosis in healthy newborn pigs. Front. Neurol. 9, 294.

Safwat, A.M., Daniel, D., 1983. Grand mal seizure after fentanyl administration. Anesthesiology 59, 78.

Sahraei, H., Rostamkhani, F., Tekieh, E., Dehghani, L., Poorazizi, E., Meamar, R., Kazemi, M., 2013. Identification of morphine accumulation in the rat embryo central nervous system: a c14-morphine administration study. Int. J. Prev. Med. 4, S222–S228.

Salgado, R.A., Jorens, P.G., Baar, I., Cras, P., Hans, G., Parizel, P.M., 2010. Methadone-induced toxic leukoencephalopathy: Mr imaging and MR proton spectroscopy findings. AJNR Am. J. Neuroradiol. 31, 565–566.

Sanchez, V., Carpenter, M.D., Yohn, N.L., Blendy, J.A., 2016. Long-lasting effects of adolescent oxycodone exposure on reward-related behavior and gene expression in mice. Psychopharmacology (Berl) 233, 3991–4002.

Schepis, T.S., Wastila, L., Ammerman, B., McCabe, V.V., McCabe, S.E., 2020. Prescription opioid misuse motives in US older adults. Pain Med. 21, 2237–2243.

Schwarz, J.M., Bilbo, S.D., 2013. Adolescent morphine exposure affects long-term microglial function and later-life relapse liability in a model of addiction. J. Neurosci. 33, 961–971.

Schwarz, J.M., Hutchinson, M.R., Bilbo, S.D., 2011. Early-life experience decreases drug-induced reinstatement of morphine CPP in adulthood via microglial-specific epigenetic programming of anti-inflammatory IL-10 expression. J. Neurosci. 31, 17835–17847.

Scott, J.C., Sarnquist, F.H., 1985. Seizure-like movements during a fentanyl infusion with absence of seizure activity in a simultaneous EEG recording. Anesthesiology 62, 812–814.

Sebel, P.S., Bovill, J.G., Wauquier, A., Rog, P., 1981. Effects of high-dose fentanyl anesthesia on the electroencephalogram. Anesthesiology 55, 203–211.

Semple, B.D., Blomgren, K., Gimlin, K., Ferriero, D.M., Noble-Haeusslein, L.J., 2013. Brain development in rodents and humans: identifying benchmarks of maturation and vulnerability to injury across species. Prog. Neurobiol. 106-107, 1–16.

Shafaroodi, H., Samini, M., Moezi, L., Homayoun, H., Sadeghipour, H., Tavakoli, S., Hajrasouliha, A.R., Dehpour, A.R., 2004. The interaction of cannabinoids and opioids on pentylenetetrazole-induced seizure threshold in mice. Neuropharmacology 47, 390–400.

Shiraishi, K., Takeda, Y., Masui, K., Taninishi, H., Sasaki, T., Danura, T., Morita, K., 2011. Effect of fentanyl on ischemic depolarization and ischemic neuronal damage of hippocampal CA1 in the gerbil. J. Anesth. 25, 540–548.

Singh, R., Saini, M., 2015. Toxic leucoencephalopathy after 'chasing the dragon'. Singapore Med. J. 56, e102–e104.

Sklair-Tavron, L., Shi, W.X., Lane, S.B., Harris, H.W., Bunney, B.S., Nestler, E.J., 1996. Chronic morphine induces visible changes in the morphology of mesolimbic dopamine neurons. Proc. Natl. Acad. Sci. U. S. A. 93, 11202–11207.

Slotkin, T.A., Weigel, S.J., Whitmore, W.L., Seidler, F.J., 1982. Maternal methadone administration: deficit in development of alpha-noradrenergic responses in developing rat brain as assessed by norepinephrine stimulation of 33Pi incorporation into phospholipids in vivo. Biochem. Pharmacol. 31, 1899–1902.
Small, J.E., Butler, P.M., Zabar, Y., Barash, J.A., 2016. Complete, bilateral hippocampal ischemia: a case series. Neurocase 22, 411–415.
Smith, N.T., Benthuysen, J.L., Bickford, R.G., Sanford, T.J., Blasco, T., Duke, P.C., Head, N., Dec-Silver, H., 1989. Seizures during opioid anesthetic induction—are they opioid-induced rigidity? Anesthesiology 71, 852–862.
Snead 3rd, O.C., 1986. Opiate-induced seizures: a study of mu and delta specific mechanisms. Exp. Neurol. 93, 348–358.
Sprung, J., Schedewie, H.K., 1992. Apparent focal motor seizure with a jacksonian march induced by fentanyl: a case report and review of the literature. J. Clin. Anesth. 4, 139–143.
Stanley, T.H., 1992. The history and development of the fentanyl series. J. Pain Symptom Manage. 7, S3–S7.
Stanley, T.H., 2014. The fentanyl story. J. Pain 15, 1215–1226.
Stiles, J., Jernigan, T.L., 2010. The basics of brain development. Neuropsychol. Rev. 20, 327–348.
Tarazi, F.I., Tomasini, E.C., Baldessarini, R.J., 1998a. Postnatal development of dopamine and serotonin transporters in rat caudate-putamen and nucleus accumbens septi. Neurosci. Lett. 254, 21–24.
Tarazi, F.I., Tomasini, E.C., Baldessarini, R.J., 1998b. Postnatal development of dopamine D4-like receptors in rat forebrain regions: comparison with D2-like receptors. Brain Res. Dev. Brain Res. 110, 227–233.
Tarazi, F.I., Tomasini, E.C., Baldessarini, R.J., 1999. Postnatal development of dopamine D1-like receptors in rat cortical and striatolimbic brain regions: an autoradiographic study. Dev. Neurosci. 21, 43–49.
Teicher, M.H., Andersen, S.L., Hostetter Jr., J.C., 1995. Evidence for dopamine receptor pruning between adolescence and adulthood in striatum but not nucleus accumbens. Brain Res. Dev. Brain Res. 89, 167–172.
Tempel, A., Yiang, J., Badheer, R., 1994. Alterations in opioid peptides and the cAMP system in brains of morphine addicted newborns. Regulatory Peptides Suppl. 1, S281–S282.
Tempelhoff, R., Modica, P.A., Bernardo, K.L., Edwards, I., 1992. Fentanyl-induced electrocorticographic seizures in patients with complex partial epilepsy. J. Neurosurg. 77, 201–208.
Tommasino, C., Maekawa, T., Shapiro, H.M., Keifer-Goodman, J., Kohlenberger, R.W., 1984. Fentanyl-induced seizures activate subcortical brain metabolism. Anesthesiology 60, 283–290.
Tormoehlen, L.M., 2013. Toxic leukoencephalopathies. Psychiatr. Clin. North Am. 36, 277–292.
U.S. Food And Drug Administration, 2015. FDA Drug Safety Communication: FDA Has Reviewed Possible Risks of Pain Medicine Use During Pregnancy.
Uddayasankar, U., Lee, C., Oleschuk, C., Eschun, G., Ariano, R.E., 2018. The pharmacokinetics and pharmacodynamics of Carfentanil after recreational exposure: A case report. Pharmacotherapy 38, e41–e45.
Urca, G., Frenk, H., Liebeskind, J.C., Taylor, A.N., 1977. Morphine and enkephalin: analgesic and epileptic properties. Science 197, 83–86.
Varshneya, N.B., Walentiny, D.M., Moisa, L.T., Walker, T.D., Akinfiresoye, L.R., Beardsley, P.M., 2019. Opioid-like antinociceptive and locomotor effects of emerging fentanyl-related substances. Neuropharmacology 151, 171–179.

Varshneya, N.B., Walentiny, D.M., Moisa, L.T., Walker, T.D., Akinfiresoye, L.R., Beardsley, P.M., 2021. Fentanyl-related substances elicit antinociception and hyperlocomotion in mice via opioid receptors. Pharmacol. Biochem. Behav. 208, 173242.
Varshneya, N.B., Hassanien, S.H., Holt, M.C., Stevens, D.L., Layle, N.K., Bassman, J.R., Iula, D.M., Beardsley, P.M., 2022. Respiratory depressant effects of fentanyl analogs are opioid receptor-mediated. Biochem. Pharmacol. 195, 114805.
Vaselbehagh, M., Sadegh, M., Karami, H., Babaie, S., Sakhaie, M.H., 2021. Coenzyme Q10 modulates apoptotic effects of chronic administration of methadone on NMRI mouse hippocampus. Cell J. 23, 538–543.
Vassoler, F.M., Byrnes, E.M., 2021. Transgenerational effects on anxiety-like behavior following adolescent morphine exposure in female rats. Behav. Brain Res. 406, 113239.
Vassoler, F.M., Byrnes, E.M., Pierce, R.C., 2014. The impact of exposure to addictive drugs on future generations: physiological and behavioral effects. Neuropharmacology 76 (Pt B), 269–275.
Vassoler, F.M., Wright, S.J., Byrnes, E.M., 2016. Exposure to opiates in female adolescents alters mu opiate receptor expression and increases the rewarding effects of morphine in future offspring. Neuropharmacology 103, 112–121.
Vassoler, F.M., Oliver, D.J., Wyse, C., Blau, A., Shtutman, M., Turner, J.R., Byrnes, E.M., 2017. Transgenerational attenuation of opioid self-administration as a consequence of adolescent morphine exposure. Neuropharmacology 113, 271–280.
Vassoler, F.M., Toorie, A.M., Byrnes, E.M., 2019. Increased cocaine reward in offspring of females exposed to morphine during adolescence. Psychopharmacology (Berl) 236, 1261–1272.
Vathy, I., Slamberova, R., Rimanoczy, A., Riley, M.A., Bar, N., 2003. Autoradiographic evidence that prenatal morphine exposure sex-dependently alters mu-opioid receptor densities in brain regions that are involved in the control of drug abuse and other motivated behaviors. Prog. Neuropsychopharmacol. Biol. Psychiatry 27, 381–393.
Vestal-Laborde, A.A., Eschenroeder, A.C., Bigbee, J.W., Robinson, S.E., Sato-Bigbee, C., 2014. The opioid system and brain development: effects of methadone on the oligodendrocyte lineage and the early stages of myelination. Dev. Neurosci. 36, 409–421.
Walker, M.L., Patel, K., Li, T., Kassir, M., 2021. Opioid-associated amnestic syndrome. Cureus 13, e20056.
Wheaton, T., Toll, B.J., Breznak, K., Da-Silva, S., Melvin, J., Misra, A., Hwang, S.W., 2019. Opioid-induced toxic leukoencephalopathy: A case report and review of the literature. Heliyon 5, e03005.
Wilde, M., Pichini, S., Pacifici, R., Tagliabracci, A., Busardo, F.P., Auwarter, V., Solimini, R., 2019. Metabolic pathways and potencies of new fentanyl analogs. Front. Pharmacol. 10, 238.
Wolters, E.C., Van Wijngaarden, G.K., Stam, F.C., Rengelink, H., Lousberg, R.J., Schipper, M.E., Verbeeten, B., 1982. Leucoencephalopathy after inhaling "heroin" pyrolysate. Lancet 2, 1233–1237.
Wu, V.W., Mo, Q., Yabe, T., Schwartz, J.P., Robinson, S.E., 2001. Perinatal opioids reduce striatal nerve growth factor content in rat striatum. Eur. J. Pharmacol. 414, 211–214.
Yaksh, T., Wallace, M., 2017. Opioids, analgesia, and pain management. In: Brunton, L.L., Hilal-Dandan, R., Knollmann, B.C. (Eds.), Goodman & Gilman's: the pharmacological basis of therapeutics, 13e, 13 ed. McGraw-Hill Education.
Yang, S.N., Huang, L.T., Wang, C.L., Chen, W.F., Yang, C.H., Lin, S.Z., Lai, M.C., Chen, S.J., Tao, P.L., 2003. Prenatal administration of morphine decreases CREBSerine-133 phosphorylation and synaptic plasticity range mediated by glutamatergic transmission in the hippocampal CA1 area of cognitive-deficient rat offspring. Hippocampus 13, 915–921.

Yang, S.N., Liu, C.A., Chung, M.Y., Huang, H.C., Yeh, G.C., Wong, C.S., Lin, W.W., Yang, C.H., Tao, P.L., 2006. Alterations of postsynaptic density proteins in the hippocampus of rat offspring from the morphine-addicted mother: beneficial effect of dextromethorphan. Hippocampus 16, 521–530.
Yazdy, M.M., Mitchell, A.A., Tinker, S.C., Parker, S.E., Werler, M.M., 2013. Periconceptional use of opioids and the risk of neural tube defects. Obstet. Gynecol. 122, 838–844.
Zagon, I.S., Mclaughlin, P.J., 1978. Perinatal methadone exposure and brain development: a biochemical study. J. Neurochem. 31, 49–54.
Zagon, I.S., Mclaughlin, P.J., 1979. Motor activity and learning ability in rats perinatally exposed to methadone. NIDA Res. Monogr. 27, 121–127.
Zagon, I.S., Mclaughlin, P.J., 1984. Perinatal exposure to methadone alters sensitivity to drugs in adult rats. Neurobehav. Toxicol. Teratol. 6, 319–323.
Zagon, I.S., Mclaughlin, P.J., Thompson, C.I., 1979. Learning ability in adult female rats perinatally exposed to methadone. Pharmacol. Biochem. Behav. 10, 889–894.
Zante, B., Hammel, E.M., Lauber, A., Schefold, J.C., 2018. 'Chasing the dragon' on the intensive care unit. Anaesthesiol. Intensive Ther. 50, 243.
Zawilska, J.B., Kuczynska, K., Kosmal, W., Markiewicz, K., Adamowicz, P., 2021. Carfentanil - from an animal anesthetic to a deadly illicit drug. Forensic Sci. Int. 320, 110715.
Zieglgansberger, W., French, E.D., Siggins, G.R., Bloom, F.E., 1979. Opioid peptides may excite hippocampal pyramidal neurons by inhibiting adjacent inhibitory interneurons. Science 205, 415–417.
Zmitrovich, A.C., Brake, S.C., Liu, P.Y., Hamowy, A.S., Hutchings, D.E., 1994. Prenatal administration of methadone in the rat: acoustic startle amplitude and the rest-activity cycle at 30 days of age. Neurotoxicol. Teratol. 16, 251–255.
Zuba, J.R., Greenberg, M., 2019. Use of naltrexone and atipamezole in emergency response to human exposure to ultra-potent opioids and alpha-2 agonists in zoo and wildlife medicine. In: Miller, R.E., Lamberski, N., Calle, P.P. (Eds.), Fowler's Zoo and Wild Animal Medicine Current Therapy. Elsevier, St. Louis, Missouri.

CHAPTER FIVE

Phencyclidine (PCP)-induced neurotoxicity and behavioral deficits

Cheng Wang[a,*], Shuliang Liu[a], Leah E. Latham[a], Fang Liu[a], Tucker A. Patterson[b], and William Slikker[b,⋆]
[a]Division of Neurotoxicology, National Center for Toxicological Research/FDA, Jefferson, AR, United States
[b]National Center for Toxicological Research/FDA, Jefferson, AR, United States
*Corresponding author: e-mail address: cheng.wang@fda.hhs.gov

Contents

1. Introduction

Phencyclidine (PCP) is a drug of abuse that interacts with many neurotransmitter systems and has the unique ability to mimic both positive and negative symptoms of schizophrenia (Cohen et al., 1962; Javitt, 1987; Javitt and Zukin, 1991). Its primary action appears to be blockade of the *N*-methyl-D-aspartate (NMDA) receptor complex (Javitt and Zukin, 1991; Lipton and Nakanishi, 1999; Rakic and Komuro, 1995). Blockade of NMDA receptors can result in neural cell damage in the developing brain (Ikonomidou et al., 1999; Scallet et al., 2004; Slikker et al., 2007; Young et al., 2005). Acute administration of PCP and other antagonists of the NMDA-type glutamate receptor has been reported to produce necrotic vacuolization of neurons associated with increases in HSP70 and glial

⋆ Retired.

Advances in Neurotoxicology, Volume 8
ISSN 2468-7480
https://doi.org/10.1016/bs.ant.2022.04.005

fibrillary acidic protein (GFAP) in the retrosplenial/posterior cingulate cortex, as well as in certain corticolimbic regions (Fix et al., 1993; Olney et al., 1989; Sharp et al., 1992). Long-term/repeated PCP administration leads to persistent schizophrenia symptoms including, among others, impairment in thinking, inability to distinguish reality from fantasy, and a loss of emotion or flattened affect (Ellison, 1995). Atrophy in the prefrontal and cingulate cortices, and hippocampal areas of the brain as demonstrated by neuroimaging studies, may underlie the behavioral deficits observed in PCP addicts (Micallef et al., 2003; Wang et al., 2004). Currently, the effects of chronic PCP administration and the underlying mechanisms are not well identified. Although PCP abuse has waned over recent years, there remains the need to understand the neurobiological effects of repeated administration. In fact, repeated ingestion has been reported to result in long-lasting behavioral changes including both positive (hallucinations, paranoia, and delusions) and negative symptoms (social withdrawal, flattened affect, and poverty of speech) of schizophrenia (Carlin et al., 1979; Cosgrove and Newell, 1991). Alternatively, ketamine is an FDA approved treatment for depression.

The mature/adult CNS appears more resistant to the neurotoxic effects of NMDA antagonists than the developing brain (Slikker et al., 2007; Wang et al., 2020). Importantly, these studies suggest that activation of NMDA-type glutamate receptors is essential for biological/physiological processes , e.g., synaptogenesis, differentiation and viability during development, and that disturbance of NMDA receptor function decreases synaptogenesis and neuronal cell-cell communication (Lipton and Nakanishi, 1999; Rakic and Komuro, 1995; Wang et al., 2020).

2. Experimental evidence for PCP-induced neurotoxicity

Previous data has shown that long-term or repeated exposure to NMDA receptor antagonists during development may cause acute widespread apoptotic neuronal damage in the cerebral cortex and subcortical structures, and long-lasting behavioral deficits/impairments involving memory and learning functions (Jevtovic-Todorovic et al., 2003; Paule et al., 2011; Shen et al., 2013). Also, critical psychological processes, such as motivation, inhibition, sequence learning and counting that are associated with the networks/circuits connecting the frontal cortex and the striatum, are disrupted after early-life exposure to NMDA receptor-based anesthesia (Paule et al., 2011).

Acute exposure to PCP results in vacuolization, neural cell swelling and reactive gliosis indicative of necrotic cell death (Fix et al., 1994, 1995; O'Callaghan, 1994). Chronic PCP exposure leads to cell death in certain regions of the brain , exhibiting apoptotic attributes characterized as a condensed or fragmented nucleus (at both morphological and biochemical levels) and reduced nerve cell volume. It is proposed that the characteristics of excitotoxic neural damage depend on the degree of brain maturity and glutamate receptor subtype along a morphological gradient ranging from apoptosis to necrosis (Portera-Cailliau et al., 1997).

Pathological data from clinical mental illness studies indicate that NMDA antagonists and/or PCP-induced neurotoxicity and behavioral changes involve disturbances in glutamatergic neurons in the most sensitive brain areas, e.g., frontal cortex. These pathophysiological changes in the frontal cortex may significantly affect the function of subcortical structures, such as striatum and/or the nucleus accumbens (Albin et al., 1992; Wang et al., 2020), rich dopaminergic projection areas thought to be critical in mental illness (Graybiel, 1990).

Accumulated data have revealed that these pathological changes arise as a consequence of structural and behavioral deficits during development (Wang et al., 2020). Previous in vivo studies demonstrated that exposure to PCP during an early developmental stage, characterized by a high level of synaptogenesis, predictably induces an apoptotic neurodegenerative reaction that might damage or even delete large numbers of neurons from the frontal cortex of the developing brain (Wang et al., 2001). Meanwhile, previous studies have indicated that repeated exposures of perinatal animals to PCP lead to selective apoptotic neuronal damage in the frontal cortex, accompanied by behavioral abnormalities including diminished baseline prepulse inhibition (PPI), locomotor sensitization and impaired spatial alternation (McInnis et al., 2002; Wang et al., 2001, 2003).

3. Potential basic mechanisms underlying PCP-induced neurotoxicity

The application of advanced approaches in both in vivo and in vitro models allowed for the dissection/evaluation of cellular and molecular mechanisms that underlie NMDA antagonists, such as PCP/ketamine-induced neurotoxicity. As mentioned, both necrotic and apoptotic mechanisms may be involved in NMDA antagonist-induced neurotoxicity.

Previous in vivo/in vitro studies (Slikker et al., 2007; Wang et al., 2005, 2006) have provided systematic/critical information for understanding that NMDA antagonists could produce neurotoxicity and/or neural damage by mechanisms involving compensatory up-regulation of NMDA-type glutamate receptors. It is proposed that continuously blocking/inhibiting NMDA-type glutamate receptors on immature neurons by NMDA receptor antagonists , e.g., PCP or ketamine, may produce compensatory upregulation of the NMDA receptor (Wang et al., 2006). Activation of these upregulated NMDA receptors by endogenous glutamate, after ketamine/PCP washout, can elevate the calcium influx or intracellular free calcium levels that exceeds the buffering capacity of neurons and interferes with electron transport in a manner that results in elevated production of reactive oxygen species and subsequent increased neuronal damage/neurodegeneration (Liu et al., 2013). In fact, data from organotypic cultures support the hypothesis that chronic PCP exposure could cause a selective loss of neurons in the frontal cortex (Johnson et al., 1998; Wang et al., 2003). By using TUNEL-staining and DNA fragmentation, previous experiments have shown that PCP causes a significant increase in TUNEL-positive cells in the frontal cortex (following PCP washout), but not in the nucleus accumbens (Johnson et al., 1998). Consistent with TUNEL data, PCP exposure results in enhanced cortical apoptosis as measured by an ELISA for histone-associated DNA fragmentation. An increase in the proapoptotic protein, Bax, was shown using Western blot analysis (Wang et al., 2000, 2003, 2004). It was also shown that prolonged NMDA antagonist (e.g., ketamine)-induced neurodegeneration in monkey frontal cortical cultures appears to be both apoptotic and necrotic in nature (Wang et al., 2006, 2017). NMDA antagonist exposure produces an apparent increase in NMDA receptor NR1 subunit expression (Hanania et al., 1999; Johnson et al., 1998; Shi et al., 2010; Wang et al., 2006). Meanwhile, NMDA receptor antagonist (e.g., ketamine)-induced neurotoxicity has been demonstrated to be associated with an increased Ca^{2+} influx/calcium overload, which is thought to be caused by glutamatergic stimulation of up-regulated NMDA receptors (Liu et al., 2013; Wang et al., 2006, 2017).

Neural cell adhesion molecule (NCAM) has a critical role in processing/regulating neuronal plasticity. Particularly, the polysialic acid form of NCAM (PSA-NCAM) is believed to play important roles in regulating/forming connections between neuronal cells (Butler et al., 1999; Edelman, 1986). The regulation of striatal PSA-NCAM expression by neuronal activity is not presently understood. The cell surface expression of PSA-NCAM is calcium-dependent (Kiss et al., 1993, 1994), possibly because of the need for released neurotransmitter(s). It is proposed that loss of cortical neurons will not only affect intracortical communication but could also affect cortical-subcortical communication and would have an adverse effect on development of brain regions that receive cortical input (Smith and Bolam, 1990). Previous data show that perinatal PCP exposure leads to selective loss of PSA-NCAM immunoreactivity in the striatum (Wang et al., 2020). It was shown that PCP exposure (in cortico-striatal organotypic culture models) decreases PSA-NCAM expression in the striatal-nucleus accumbens (Wang et al., 2020). Because PSA expression may be regulated by glutamate, PCP-induced cortical damage (and presumed loss of glutamatergic input) can result in a loss of PSA-NCAM and subsequent disorganization of striatal neural networks. In fact, using a pre-embedding immuno-peroxidase technique [via electron microscopy (EM)], intense PSA immunoreactivity can be observed on the synaptic contacts of both pre- and post-synaptic membrane surfaces in the control section, and PSA expression is reduced in the PCP-treated section (Wang et al., 2020). These data support the hypothesis that PSA is localized to synaptic contacts and regulation of PSA-NCAM expression in the striatal-nucleus accumbens complex is activity dependent.

PCP-induced apoptosis in the frontal cortex may directly affect the development and function of the striatum-nucleus accumbens complex as this area is known to receive substantial glutamatergic input from the frontal cortex (Albin et al., 1992). Previous data shows that exposure to PCP on postnatal days (PND) 7, 9 and 11 results in neurodegenerative damage and/or deletion of a significant number of neurons from the frontal cortex of the developing brain (Wang et al., 2001). Importantly, this regimen results in behavioral changes including diminished baseline prepulse inhibition (PPI) of acoustic startle on PND 24–28, a sensitized locomotor response to PCP challenge on PND 42, and a deficit in acquisition of a spatial alternation task on PND 33–70 (McInnis et al., 2002; Wang et al., 2001, 2003). Thus, the loss of cortical neurons and a decrease of PSA-NCAM expression in the striatal-nucleus accumbens complex could be a basic mechanism for PCP induced neurotoxicity, and in turn, could result in the changes in striatal synaptogenesis during development. The striatal nucleus-accumbens complex is abundantly (dopaminergic) innervated and is thought to be associated with the development of mental disorders (Mavridis, 2015; Whittaker et al., 2018). Also, observed pathophysiological changes are believed to link to mental/behavioral abnormalities (Hollis, 2003; Wang et al., 2001; Watt et al., 2017) and addictive behaviors (McInnis et al., 2002; Scofield et al., 2016; Wang et al., 2003). In general, PSA-NCAM activity and modification of PSA-NCAM activity is critical to the regulation of synaptic protein expression. The dysregulation/dysfunction of subcortical synaptogenesis and cortico-striatal interaction/communication underlies the behavioral deficits and schizophrenia-like pathology produced by repeated PCP exposure in the rodent.

4. Prevention/improvement of PCP-induced neurotoxicity

As a tool compound, PCP could potentially be applied to understand the neurobiological effects in addiction and highlight its roles in developmental abnormalities of schizophrenia. In the context of the basic mechanisms that behavioral abnormalities/mental disorders and cortical neuronal damage/apoptosis are associated with alterations of NMDA receptor subunit mRNA and/or protein expression, the available neuroprotective treatments (both in vivo and in vitro) can intervene by interaction with the NMDA-type glutamate receptor system. Previous in situ hybridization and immunocytochemistry data have demonstrated that repeated exposure to the NMDA antagonists, e.g., PCP, results in an upregulation of the NR1 subunit of the NMDA receptor in the forebrain structure (compensatory adaptation by the receptor) and appears critical to subsequent neuronal loss (Johnson et al., 1998; Wang et al., 1999, 2000). The NMDA receptor is a ligand-activated ion channel primarily composed of three families of NMDA receptor subunits: NR1, NR2 (A-D) and NR3 (A and B) (Ishii et al., 1993; Moriyoshi et al., 1991; Paupard et al., 1997; Wong et al., 2002; Zukin and Bennett, 1995). The NR1 subunit is an essential/fundamental subunit necessary for NMDA channel function. Therefore, it is practically applicable to target the NR1 subunit to block NMDA receptor NR1 synthesis and effectively diminish/prevent cell death (apoptotic) cascades caused by PCP (Wang et al., 2006, 2017).

It is also important to consider synthesized drugs such as M40403 [or M40404, its inactive analog (Salvemini et al., 1999)] and SN50. M40403 is a non-peptidyl superoxide mimetic, and SN50 is a peptide inhibitor of NF-*k*B protein translocation (McInnis et al., 2002). These compounds seem to have the potential for being used to block/prevent PCP-induced cortical neuronal damage (McInnis et al., 2002). Previous data indicated that nuclear translocation of NF-*k*B is associated with elevation of Bax (proapoptotic protein) and DNA fragmentation, and both can be blocked by either M40403 or SN50. These studies revealed that superoxide anions are involved in the apoptotic pathway that is related to NF-*k*B transcription factors and is sensitive to the redox status of the surrounding milieu. As mentioned, repeated PCP administration may result in long-term behavioral deficits and these behavioral disturbances could be associated with neuronal apoptosis in the forebrain. Thus, the ability of M40403 and SN50 to block/prevent both the biochemical and behavioral sequelae of repeated PCP exposure supports the correlation (basic mechanism) between cortical neuronal loss and behavioral/biochemical changes/abnormalities (Wang et al., 2003). In addition, the effectiveness of M40403 and SN50 in this cascade/paradigm indicates a potential therapeutic approach for PCP addicts or those at risk for schizophrenia. Thus, based on the discussed (basic) mechanisms that cortical damage and/or disruption of cortico-striatal communication and dysfunction of subcortical synaptogenesis underlies the behavioral deficits or addictive states, some target molecules/agents shown to be protective in other models of neurotoxicity could be used to prevent or ameliorate any of the toxic effects associated with repeated PCP administration.

5. Summary

Accumulated evidence reveals the basic neurotoxic mechanisms of PCP administration and the relationship between dysregulation of NMDA receptors and neuronal loss in forebrain during development. Cortical neuronal loss may significantly impact the function of subcortical structures, such as the striatal nucleus-accumbens complex, which may serve as a critical regulatory center integrating the functions of the basal ganglia and the limbic system. These subcortical regions have been implicated in behavioral disturbances observed in PCP abusers and individuals at risk for schizophrenia.

Acknowledgments

This book chapter reflects the views of the authors and does not necessarily reflect those of the U.S. Food and Drug Administration.

References

Albin, R.L., Makowiec, R.L., Hollingsworth, Z.R., Dure, L.S.T., Penney, J.B., Young, A.B., 1992. Excitatory amino acid binding sites in the basal ganglia of the rat: a quantitative autoradiographic study. Neuroscience 46, 35–48.

Butler, A.K., Uryu, K., Rougon, G., Chesselet, M.F., 1999. N-methyl-D-aspartate receptor blockade affects polysialylated neural cell adhesion molecule expression and synaptic density during striatal development. Neuroscience 89, 1169–1181.

Carlin, A.S., Grant, I., Adams, K.M., Reed, R., 1979. Is phencyclidine (PCP) abuse associated with organic mental impairment? Am. J. Drug Alcohol Abuse 6, 273–281.

Cohen, B.C., Rosenbaum, G., Luby, E.D., Gottlieb, J.S., 1962. Comparison of phecyclidine hydrochloride with other drugs. Stimulation of schizophrenic performance with phencyclidine hydrochloride, lysergic acid diethylamide and amobarbital sodium. Arch. Gen. Psychiatry, 395–401.

Cosgrove, J., Newell, T.G., 1991. Recovery of neuropsychological functions during reduction in use of phencyclidine. J. Clin. Psychol. 47, 159–169.

Edelman, G.M., 1986. Cell adhesion molecules in the regulation of animal form and tissue pattern. Annu. Rev. Cell Biol. 2, 81–116.

Ellison, G., 1995. The N-methyl-D-aspartate antagonists phencyclidine, ketamine and dizocilpine as both behavioral and anatomical models of the dementias. Brain Res. Brain Res. Rev. 20, 250–267.

Fix, A.S., Horn, J.W., Truex, L.L., Smith, R.A., Gomez, E., 1994. Neuronal vacuole formation in the rat posterior cingulate/retrosplenial cortex after treatment with the N-methyl-D-aspartate (NMDA) antagonist MK-801 (dizocilpine maleate). Acta Neuropathol. 88, 511–519.

Fix, A.S., Horn, J.W., Wightman, K.A., Johnson, C.A., Long, G.G., Storts, R.W., Farber, N., Wozniak, D.F., Olney, J.W., 1993. Neuronal vacuolization and necrosis induced by the noncompetitive N-methyl-D-aspartate (NMDA) antagonist MK(+)801 (dizocilpine maleate): a light and electron microscopic evaluation of the rat retrosplenial cortex. Exp. Neurol. 123, 204–215.

Fix, A.S., Wightman, K.A., O'callaghan, J.P., 1995. Reactive gliosis induced by MK-801 in the rat posterior cingulate/retrosplenial cortex: GFAP evaluation by sandwich ELISA and immunocytochemistry. Neurotoxicology 16, 229–237.

Graybiel, A.M., 1990. Neurotransmitters and neuromodulators in the basal ganglia. Trends Neurosci. 13, 244–254.

Hanania, T., Hillman, G.R., Johnson, K.M., 1999. Augmentation of locomotor activity by chronic phencyclidine is associated with an increase in striatal NMDA receptor function and an upregulation of the NR1 receptor subunit. Synapse 31, 229–239.

Hollis, C., 2003. Developmental precursors of child- and adolescent-onset schizophrenia and affective psychoses: diagnostic specificity and continuity with symptom dimensions. Br. J. Psychiatry 182, 37–44.

Ikonomidou, C., Bosch, F., Miksa, M., Bittigau, P., Vockler, J., Dikranian, K., Tenkova, T.I., Stefovska, V., Turski, L., Olney, J.W., 1999. Blockade of NMDA receptors and apoptotic neurodegeneration in the developing brain. Science 283, 70–74.

Ishii, T., Moriyoshi, K., Sugihara, H., Sakurada, K., Kadotani, H., Yokoi, M., Akazawa, C., Shigemoto, R., Mizuno, N., Masu, M., et al., 1993. Molecular characterization of the family of the N-methyl-D-aspartate receptor subunits. J. Biol. Chem. 268, 2836–2843.

Javitt, D.C., 1987. Negative schizophrenic symptomatology and the PCP (phencyclidine) model of schizophrenia. Hillside J. Clin. Psychiatry 9, 12–35.

Javitt, D.C., Zukin, S.R., 1991. Recent advances in the phencyclidine model of schizophrenia. Am. J. Psychiatry 148, 1301–1308.

Jevtovic-Todorovic, V., Hartman, R.E., Izumi, Y., Benshoff, N.D., Dikranian, K., Zorumski, C.F., Olney, J.W., Wozniak, D.F., 2003. Early exposure to common anesthetic agents causes widespread neurodegeneration in the developing rat brain and persistent learning deficits. J. Neurosci. 23, 876–882.

Johnson, K.M., Phillips, M., Wang, C., Kevetter, G.A., 1998. Chronic phencyclidine induces behavioral sensitization and apoptotic cell death in the olfactory and piriform cortex. J. Neurosci. Res. 52, 709–722.

Kiss, J.Z., Wang, C., Rougon, G., 1993. Nerve-dependent expression of high polysialic acid neural cell adhesion molecule in neurohypophysial astrocytes of adult rats. Neuroscience 53, 213–221.

Kiss, J.Z., Wang, C., Olive, S., Rougon, G., Lang, J., Baetens, D., Harry, D., Pralong, W.F., 1994. Activity-dependent mobilization of the adhesion molecule polysialic NCAM to the cell surface of neurons and endocrine cells. EMBO J. 13, 5284–5292.

Lipton, S.A., Nakanishi, N., 1999. Shakespeare in love- -with NMDA receptors? Nat. Med. 5, 270–271.

Liu, F., Patterson, T.A., Sadovova, N., Zhang, X., Liu, S., Zou, X., Hanig, J.P., Paule, M.G., Slikker Jr., W., Wang, C., 2013. Ketamine-induced neuronal damage and altered N-methyl-D-aspartate receptor function in rat primary forebrain culture. Toxicol. Sci. 131, 548–557.

Mavridis, I., 2015. The role of the nucleus accumbens in psychiatric disorders. Psychiatriki 25, 282–294.

McInnis, J., Wang, C., Anastasio, N., Hultman, M., Ye, Y., Salvemini, D., Johnson, K.M., 2002. The role of superoxide and nuclear factor-kappaB signaling in N-methyl-D-aspartate-induced necrosis and apoptosis. J. Pharmacol. Exp. Ther. 301, 478–487.

Micallef, J., Tardieu, S., Gentile, S., Fakra, E., Jouve, E., Sambuc, R., Blin, O., 2003. Effects of a subanaesthetic dose of ketamine on emotional and behavioral state in healthy subjects. Neurophysiol. Clin. 33, 138–147.

Moriyoshi, K., Masu, M., Ishii, T., Shigemoto, R., Mizuno, N., Nakanishi, S., 1991. Molecular cloning and characterization of the rat NMDA receptor. Nature 354, 31–37.

O'Callaghan, J.P., 1994. Biochemical analysis of glial fibrillary acidic protein as a quantitative approach to neurotoxicity assessment: advantages, disadvantages and application to the assessment of NMDA receptor antagonist-induced neurotoxicity. Psychopharmacol. Bull. 30, 549–554.

Olney, J.W., Labruyere, J., Price, M.T., 1989. Pathological changes induced in cerebrocortical neurons by phencyclidine and related drugs. Science 244, 1360–1362.

Paule, M.G., Li, M., Allen, R.R., Liu, F., Zou, X., Hotchkiss, C., Hanig, J.P., Patterson, T.A., Slikker Jr., W., Wang, C., 2011. Ketamine anesthesia during the first week of life can cause long-lasting cognitive deficits in rhesus monkeys. Neurotoxicol. Teratol. 33, 220–230.

Paupard, M.C., Friedman, L.K., Zukin, R.S., 1997. Developmental regulation and cell-specific expression of N-methyl-D-aspartate receptor splice variants in rat hippocampus. Neuroscience 79, 399–409.

Portera-Cailliau, C., Price, D.L., Martin, L.J., 1997. Non-NMDA and NMDA receptor-mediated excitotoxic neuronal deaths in adult brain are morphologically distinct: further evidence for an apoptosis-necrosis continuum. J. Comp. Neurol. 378, 88–104.

Rakic, P., Komuro, H., 1995. The role of receptor/channel activity in neuronal cell migration. J. Neurobiol. 26, 299–315.

Salvemini, D., Wang, Z.Q., Zweier, J.L., Samouilov, A., Macarthur, H., Misko, T.P., Currie, M.G., Cuzzocrea, S., Sikorski, J.A., Riley, D.P., 1999. A nonpeptidyl mimic of superoxide dismutase with therapeutic activity in rats. Science 286, 304–306.

Scallet, A.C., Schmued, L.C., Slikker Jr., W., Grunberg, N., Faustino, P.J., Davis, H., Lester, D., Pine, P.S., Sistare, F., Hanig, J.P., 2004. Developmental neurotoxicity of ketamine: morphometric confirmation, exposure parameters, and multiple fluorescent labeling of apoptotic neurons. Toxicol. Sci. 81, 364–370.
Scofield, M.D., Heinsbroek, J.A., Gipson, C.D., Kupchik, Y.M., Spencer, S., Smith, A.C., Roberts-Wolfe, D., Kalivas, P.W., 2016. The nucleus Accumbens: mechanisms of addiction across drug classes reflect the importance of glutamate homeostasis. Pharmacol. Rev. 68, 816–871.
Sharp, F.R., Butman, M., Wang, S., Koistinaho, J., Graham, S.H., Sagar, S.M., Noble, L., Berger, P., Longo, F.M., 1992. Haloperidol prevents induction of the hsp70 heat shock gene in neurons injured by phencyclidine (PCP), MK801, and ketamine. J. Neurosci. Res. 33, 605–616.
Shen, X., Dong, Y., Xu, Z., Wang, H., Miao, C., Soriano, S.G., Sun, D., Baxter, M.G., Zhang, Y., Xie, Z., 2013. Selective anesthesia-induced neuroinflammation in developing mouse brain and cognitive impairment. Anesthesiology 118, 502–515.
Shi, Q., Guo, L., Patterson, T.A., Dial, S., Li, Q., Sadovova, N., Zhang, X., Hanig, J.P., Paule, M.G., Slikker Jr., W., Wang, C., 2010. Gene expression profiling in the developing rat brain exposed to ketamine. Neuroscience 166, 852–863.
Slikker Jr., W., Paule, M.G., Wright, L.K., Patterson, T.A., Wang, C., 2007. Systems biology approaches for toxicology. J. Appl. Toxicol. 27, 201–217.
Smith, A.D., Bolam, J.P., 1990. The neural network of the basal ganglia as revealed by the study of synaptic connections of identified neurones. Trends Neurosci. 13, 259–265.
Wang, C., Showalter, V.M., Hillman, G.R., Johnson, K.M., 1999. Chronic phencyclidine increases NMDA receptor NR1 subunit mRNA in rat forebrain. J. Neurosci. Res. 55, 762–769.
Wang, C., Kaufmann, J.A., Sanchez-Ross, M.G., Johnson, K.M., 2000. Mechanisms of N-methyl-D-aspartate-induced apoptosis in phencyclidine-treated cultured forebrain neurons. J. Pharmacol. Exp. Ther. 294, 287–295.
Wang, C., Mcinnis, J., Ross-Sanchez, M., Shinnick-Gallagher, P., Wiley, J.L., Johnson, K.M., 2001. Long-term behavioral and neurodegenerative effects of perinatal phencyclidine administration: implications for schizophrenia. Neuroscience 107, 535–550.
Wang, C., Mcinnis, J., West, J.B., Bao, J., Anastasio, N., Guidry, J.A., Ye, Y., Salvemini, D., Johnson, K.M., 2003. Blockade of phencyclidine-induced cortical apoptosis and deficits in prepulse inhibition by M40403, a superoxide dismutase mimetic. J. Pharmacol. Exp. Ther. 304, 266–271.
Wang, C., Anastasio, N., Popov, V., Leday, A., Johnson, K.M., 2004. Blockade of N-methyl-D-aspartate receptors by phencyclidine causes the loss of corticostriatal neurons. Neuroscience 125, 473–483.
Wang, C., Sadovova, N., Fu, X., Schmued, L., Scallet, A., Hanig, J., Slikker, W., 2005. The role of the N-methyl-D-aspartate receptor in ketamine-induced apoptosis in rat forebrain culture. Neuroscience 132, 967–977.
Wang, C., Sadovova, N., Hotchkiss, C., Fu, X., Scallet, A.C., Patterson, T.A., Hanig, J., Paule, M.G., Slikker Jr., W., 2006. Blockade of N-methyl-D-aspartate receptors by ketamine produces loss of postnatal day 3 monkey frontal cortical neurons in culture. Toxicol. Sci. 91, 192–201.
Wang, C., Zhang, X., Liu, F., 2017. Application of advanced preclinical models and methods in anesthetic neurotoxicity research. Neurotoxicol. Teratol. 61, 1–6.
Wang, C., Inselman, A., Liu, S., Liu, F., 2020. Potential mechanisms for phencyclidine/ketamine-induced brain structural alterations and behavioral consequences. Neurotoxicology 76, 213–219.

Watt, M.J., Weber, M.A., Davies, S.R., Forster, G.L., 2017. Impact of juvenile chronic stress on adult cortico-accumbal function: implications for cognition and addiction. Prog. Neuropsychopharmacol. Biol. Psychiatry 79, 136–154.
Whittaker, J.R., Foley, S.F., Ackling, E., Murphy, K., Caseras, X., 2018. The functional connectivity between the nucleus Accumbens and the ventromedial prefrontal cortex as an Endophenotype for bipolar disorder. Biol. Psychiatry 84, 803–809.
Wong, H.K., Liu, X.B., Matos, M.F., Chan, S.F., Perez-Otano, I., Boysen, M., Cui, J., Nakanishi, N., Trimmer, J.S., Jones, E.G., Lipton, S.A., Sucher, N.J., 2002. Temporal and regional expression of NMDA receptor subunit NR3A in the mammalian brain. J. Comp. Neurol. 450, 303–317.
Young, C., Jevtovic-Todorovic, V., Qin, Y.Q., Tenkova, T., Wang, H., Labruyere, J., Olney, J.W., 2005. Potential of ketamine and midazolam, individually or in combination, to induce apoptotic neurodegeneration in the infant mouse brain. Br. J. Pharmacol. 146, 189–197.
Zukin, R.S., Bennett, M.V., 1995. Alternatively spliced isoforms of the NMDARI receptor subunit. Trends Neurosci. 18, 306–313.

Further reading

Fix, A.S., Long, G.G., Wozniak, D.F., Olney, J.W., 1994. Pathomorphological effects of NMDA antagonists in the rat posterior cingulate/retrosplenial cerebral cortex. Drug Dev. Res., 147–152.
Olney, J.W., Farber, N.B., 1995. NMDA antagonists as neurotherapeutic drugs, psychotogens, neurotoxins, and research tools for studying schizophrenia. Neuropsychopharmacology 13, 335–345.
Raper, J., Alvarado, M.C., Murphy, K.L., Baxter, M.G., 2015. Multiple anesthetic exposure in infant monkeys alters emotional reactivity to an acute stressor. Anesthesiology 123, 1084–1092.
Wang, C., Zheng, D., Xu, J., Lam, W., Yew, D.T., 2013. Brain damages in ketamine addicts as revealed by magnetic resonance imaging. Front Neuroanat 7, 23.

CHAPTER SIX

The mysterious beauty of ketamine: Benefits and risks

Donald R. Mattison[a,b,c,*], Abdallah Alami[b,d], Thomas J. Moore[e,f], and G. Caleb Alexander[e,g]

[a]Department of Epidemiology and Biostatistics, Arnold School of Public Health, University of South Carolina, Columbia, SC, United States
[b]McLaughlin Centre for Population Health Risk Assessment, University of Ottawa, Ottawa, ON, Canada
[c]School of Epidemiology and Public Health, University of Ottawa, Ottawa, ON, Canada
[d]School of Mathematics and Statistics, Carleton University, Ottawa, ON, Canada
[e]Center for Drug Safety and Effectiveness, Bloomberg School of Public Health, Johns Hopkins University, Baltimore, MD, United States
[f]Department of Epidemiology, Milken Institute School of Public Health, The George Washington University, Washington, DC, United States
[g]Division of General Internal Medicine, Johns Hopkins Medicine, Baltimore, MD, United States
*Corresponding author: e-mail address: mattisd@mailbox.sc.edu

Contents

1. Introduction

Ketamine, a chiral molecule with two optical isomers (S+)-ketamine and (R-)-ketamine, has been used in the US as an aesthetic with minimal risk (Lodge and Mercier, 2015) predominantly dissociative experiences with withdrawal. These psychoactive drugs are non-competitive, non-selective antagonists of the NMDA neuroreceptors. More recently, esketamine was separated from the chiral mixture containing both R-ketamine, and (S+)-ketamine' that has been licensed in the US for -treatment-resistant

Advances in Neurotoxicology, Volume 8
ISSN 2468-7480
https://doi.org/10.1016/bs.ant.2022.06.004

depression and acute suicidality (Drugs@F.D.A, 2021). While both molecules are non-competitive antagonists of NMDA receptors (Goldberg et al., 2010), the (S+)-ketamine is 2 to 4 times more potent than ketamine (Domino and Warner, 2010; Lodge and Mercier, 2015). The immediate but transient effect of these drugs is to produce anesthesia, but during anesthetic withdrawal patients experience an altered mental state that ranges from bizarre, weird feelings to frank psychosis, and hence a model for schizophrenia is produced (Domino and Warner, 2010). This altered state of consciousness usually normalizes within 2 hours, although for some individuals it can persist for much longer times, even after the drug has been cleared from the body. While (S+)-ketamine has a Risk Evaluation and Mitigation Strategy (REMS), to deal with the psychotomimetic state during withdrawal, the R-ketamine, its enantiomer does not, which does not reflect the potential for toxicity and public health damage which can be produced by the R + enantiomer.

We consider the safety of these NMDA antagonists, approved at different times and for different indications—and focus on whether the neurotoxicity of ketamine or esketamine preclude use because of variable effects in neurotoxicity, and whether risk management strategies should be implemented for this class of drugs going forward. The beauty (benefit) of these NMDA antagonists may be smaller than anticipated.

In a previous multidisciplinary review, we have concluded that there are substantial questions about both the adequacy of clinical testing and effectiveness of both ketamine and esketamine for depression (Moore et al., 2022).

2. Between PCP and ketamine

The behavioral effects of PCP as a recreational drug captured the imagination of America—drawing attention to super-human acts while under the influence of PCP or ketamine (Bertron et al., 2018). PCP was initially synthesized in late 1950s with promising clinical pharmacology and chemical properties: readily absorbed from intravenous, oral, and nasal administration with good bioavailability and central nervous system (CNS) penetration (Domino and Luby, 2012). While the brain and adipose tissues may have detectable levels of PCP lasting long, long highs have also been observed for ketamine enantiomers when used for abuse or chronically in treatment (e.g., surgery) (Bertron et al., 2018). Given the similarities in

Table 1 Fast-acting, non-selective, non-competitive NMDA receptor antagonists.

	Phencyclidine (PCP)	Ketamine (Ketalar)	Esketamine (Spravato)
Primary use	Illegal street drug	Anesthetic	Treatment-resistant depression
Other uses	Anesthetic (withdrawn)	Psychiatric disorders	
		Illegal street drug	
Route of administration	Snorted/inhaled	IV/IM	Nasal spray device
Abuse and dependence[a]	High potential	Moderate-low potential	Moderate-low potential
Time to onset	Minutes[b]	3 min[c]	20–40 min (C_{max})

[a]US Drug Enforcement Administration Controlled Substance Schedule II and III.
[b]Not systematically assessed in illicit drug use.
[c]Onset may vary by rate of IV infusion.
Abbreviations: IV, intravenous; IM, intramuscular; C_{max}, time to maximum concentration.
Adapted with permission from Moore, T.J., Alami, A., Alexander, G.C., Mattison, D.R., 2022. Safety and effectiveness of NMDA receptor antagonists for depression: A multidisciplinary review. Pharmacotherapy 42 (7) 567–579. doi: 10.1002/phar.2707. Epub ahead of print. PMID: 35665948.

either ketamine enantiomers, concerns about similar behavioral adverse effects remain a concern, especially given that repeated dosing with recreational use produces persistent positive, negative, and cognitive system clusters (Table 1). In addition to the behavioral endpoints, the Structure-Activity Relationships (SAR) models of sensitivity to the Olney lesions (with vacuolization and apoptosis) were also observed with PCP (Bertron et al., 2018), as well as prolonged psychosis—with an extension of these endpoints to ketamine.

Early work on PCP in 1950s and 1960s and its severe effects on brain neurotoxicity led to increasingly strict regulatory constraints with ketamine replacing PCP for medical use in 1970s as an anesthetic, with an emerging interest in the treatment of depression, as well as increased recreational use (Muetzelfeldt et al., 2008). Phenomenological aspects encouraging use included "melting into the surroundings," visual hallucinations, and out of body experiences, while those unappealing aspects included memory loss and decreased sociability. Concerns included long-term health effects, K-cramps, cystitis, and dependence (Muetzelfeldt et al., 2008). Note that while ketamine may have fewer adverse effects, emergent dreams and hallucinations are still reported in post-anesthetic and post-recreational periods, with schizoidal behaviors being observed in both acute and chronic users. Perhaps facilitated by the bioavailability of ketamine. Ketamine and (S+)-ketamine, as NMDA receptor antagonists, can also cause a wide range of cognitive and behavioral problems, including psychosis. While these effects are usually temporary, there are concerns that they may promote a variety of psychotomimetic symptoms, transient dissociative symptoms, and possibly cognitive impairment.

3. Ketamine and apoptosis

A consistent concern among users of PCP and Ketamine is neurotoxicity with early work by Olney (Olney et al., 1989) describing the step-wise neurotoxicity of vacuolization through apoptosis and consequent brain damage. Long-term ketamine usage has also been demonstrated to play a role in reducing the frontal grey matter volume (Morgan et al., 2014). The connection between frontal grey matter attenuation and ketamine usage duration and cumulative doses may imply a dose-dependent impact of long-term ketamine use. Research on ketamine users with substance use disorder also revealed that various regions of the brain in humans that are vulnerable to long-term ketamine injury, and displayed a diffuse impact of ketamine on the brain that may vary from other CNS drugs such as cocaine, heroin, and methamphetamine (Gill et al., 2021). Furthermore, chronic ketamine usage was linked to extensive disruption of white matter integrity and white matter networks connecting subcortical and prefrontal cortical regions, which might also explain individual variability in ketamine-induced dissociative experiences (Gastaldon et al., 2019).

4. Ketamine-induced psychosis

One study aimed at determining which brain regions are involved in NMDA receptor-mediated psychosis, using a double-blinded randomized controlled trial design (Breier et al., 1997). Healthy volunteers were given ketamine infusions when they underwent positron emission tomography (PET) scans to localize the areas of the brain that were triggered by ketamine. Investigators noted that ketamine resulted in an acute psychotic state, as well as an increase in metabolic activity in the prefrontal cortex.

5. Ketamine and schizophrenic symptoms

According to some research, ketamine can cause brain damage and mimic the symptoms of schizophrenia (Liang et al., 2015; Pomarol-Clotet et al., 2006; Becker et al., 2003; Beck, 2020). In animal studies, both PCP and ketamine produced vacuolization and apoptosis in the rat brain (Olney et al., 1989), with similarities to schizophrenia (Lodge and Mercier, 2015). In humans, one clinical trial compared the effect of ketamine use on neuroleptic-free schizophrenic patients compared to normal controls, while investigating if patients will experience an exacerbation of psychotic symptoms and cognitive impairments (Malhotra et al., 1997). Schizophrenics experienced a brief-ketamine-induced exacerbation of positive and negative symptoms with further decrements in recall and recognition memory. Auditory hallucinations and paranoia were observed in patients but not in controls. These findings suggest that ketamine is linked to a worsening of core psychotic and cognitive symptoms in schizophrenia.

A subsequent study also evaluated the effects of ketamine on healthy and schizophrenic volunteers and reported that ketamine caused a transient increase in schizophrenic symptoms in both healthy volunteers and in patients with schizophrenia (Lahti et al., 2001). The authors suggested that ketamine's putative antagonism of NMDA-sensitive glutamatergic transmission triggers schizophrenia-like psychotic symptoms in normal controls and exacerbates certain symptoms of schizophrenia in the patient group (Lahti et al., 2001).

6. Ketamine and cognitive functioning

Perceptual effects akin to dissociative states and euphoric effects similar to ethanol effects have been reported (Krystal et al., 1999). Frequent ketamine users also displayed spatial memory deficits, accompanied by reduced hippocampal activation (Morgan et al., 2014). A group of investigators evaluated the phenomenological aspects of ketamine by surveying 90 ketamine users that were grouped into three groups: frequent users, recreational and ex-users (Muetzelfeldt et al., 2008). They used structured interviews to evaluate initiation experiences, positive and negative experiences, while verifying use with hair sample analysis. Participants were asked to describe concerns related to the long-term use of ketamine and spontaneous identification of the long-term theme—impaired cognitive functioning. Participants extended these concerns by adding declines in mental and physical health and awareness of addiction as a problem. However, a recent review by Gill et al. (2021) explored the effects of ketamine on neurocognition in treatment-resistant depression (TRD) population and reported that based on five citations that met the inclusion criteria, intravenous ketamine is not associated with anti-cognitive effects (Gill et al., 2021).

7. Conclusion

A recent review (Gastaldon et al., 2019) evaluated both the safety and efficacy of esketamine, noting that "… further evidence on safety is urgently needed" in animal and human studies. The extensive off-label use of ketamine and esketamine for multiple psychiatric disorders needs to be re-evaluated from a safety, effectiveness, legal, drug advertising, and regulatory perspective. In a previous multidisciplinary review, we concluded that there are substantial questions about both the adequacy of clinical testing and the effectiveness of both ketamine and esketamine for depression (Moore et al., 2022). While Esketamine and Ketamine have been explicitly designed with a focus on increasing clinical utility, the actual safety remains uncharacterized and their risk/benefit balance remains to be characterized.

Declarations

Funding: No funding was received for the conduct of this study or for the preparation of this manuscript.

Conflicts of interest/competing interests

Availability of data: The data for this study were from publicly available documents and studies.

Code availability: Not applicable.

Authors' contributions: DRM and AA drafted the manuscript. DRM, TJM, AA, and GCA contributed to the study conception and design. DRM, TJM, and AA led the data collection and review. TJM, GCA, and AA provided a critical review of drafts and approved the final manuscript.

References

Beck, K., et al., 2020. Association of ketamine with psychiatric symptoms and implications for its therapeutic use and for understanding schizophrenia: a systematic review and meta-analysis. JAMA Netw. Open 3 (5), e204693-e204693.

Becker, A., Peters, B., Schroeder, H., Mann, T., Huether, G., Grecksch, G., 2003. Ketamine-induced changes in rat behaviour: a possible animal model of schizophrenia. Prog. Neuropsychopharmacol. Biol. Psychiatry 27 (4), 687–700.

Bertron, J.L., Seto, M., Lindsley, C.W., 2018. DARK classics in chemical neuroscience: phencyclidine (PCP). ACS Chem. Neurosci. 9. p. 2459–74-2459–74.

Breier, A., et al., 1997. Association of ketamine-induced psychosis with focal activation of the prefrontal cortex in healthy volunteers. Am. J. Psychiatry 154 (6), 805–811.

Domino, E.F., Luby, E.D., 2012. Phencyclidine/schizophrenia: one view toward the past, the other to the future. Schizophrenia Bull. 38 (5), 914–919.

Domino, E.F., Warner, D.S., 2010. Taming the ketamine tiger. Anesthesiology 113. p. 678–84-678–84.

Drugs@F.D.A, 2021. FDA-Approved Drugs [Internet]. US Food and Drug Administration. web site.

Gastaldon, C., et al., 2019. Esketamine for treatment resistant depression: a trick of smoke and mirrors? Epidemiol. Psychiatr. Sci. 29, e79.

Gill, H., et al., 2021. The effects of ketamine on cognition in treatment-resistant depression: a systematic review and priority avenues for future research. Neurosci. Biobehav. Rev. 120, 78–85.

Goldberg, M.E., et al., 2010. Pharmacodynamic profiles of ketamine (R)- and (S)- with 5-day inpatient infusion for the treatment of complex regional pain syndrome. Pain Phys. 13 (4), 379–387.

Krystal, J.H., et al., 1999. Interactive effects of subanesthetic ketamine and haloperidol in healthy humans. Psychopharmacology (Berl) 145 (2), 193–204.

Lahti, A.C., et al., 2001. Effects of ketamine in normal and schizophrenic volunteers. Neuropsychopharmacology 25 (4), 455–467.

Liang, H., et al., 2015. Ketamine users have high rates of psychosis and/or depression. J. Addict. Nurs. 26, 8–13.

Lodge, D., Mercier, M.S., 2015. Ketamine and phencyclidine: the good, the bad and the unexpected. Br. J. Pharmacol. 172. p. 4254–76-4254–76.

Malhotra, A.K., et al., 1997. Ketamine-induced exacerbation of psychotic symptoms and cognitive impairment in neuroleptic-free schizophrenics. Neuropsychopharmacology 17 (3), 141–150.

Moore, T., et al., 2022. Safety and effectiveness of NMDA receptor antagonists for depression: a multidisciplinary review. Pharmacotherapy 42 (7), 567–579 (Epub ahead of print).

Morgan, C.J.A., et al., 2014. Long-term heavy ketamine use is associated with spatial memory impairment and altered hippocampal activation. Front. Psychiatry 5.

Muetzelfeldt, L., et al., 2008. Journey through the K-hole: phenomenological aspects of ketamine use. Drug Alcohol Depend. 95. p. 219–29-219–29.

Olney, J.W., Labruyere, J., Price, M.T., 1989. Pathological changes induced in cerebrocortical neurons by phencyclidine and related drugs. Science 244. p. 1360–2-1360–2.

Pomarol-Clotet, E., et al., 2006. Psychological effects of ketamine in healthy volunteers. Phenomenological study. Br. J. Psychiatry 189, 173–179.

FSC
www.fsc.org
MIX
Papier | Fördert gute Waldnutzung
FSC® C083411

Kolibri 360 GmbH
Ferdinand-Jühlke-Straße 7
99095 Erfurt, Deutschland
produktsicherheit@kolibri360.de